Topics in Intelligent Engineering and Informatics

Volume 11

Series editors

More information about this series at http://www.springer.com/series/10188

John Vong · Insu Song

Emerging Technologies for Emerging Markets

Springer

John Vong
Singapore Management University
Singapore
Singapore

Insu Song
James Cook University
Singapore
Singapore

ISSN 2193-9411 ISSN 2193-942X (electronic)
Topics in Intelligent Engineering and Informatics
ISBN 978-981-10-1369-0 ISBN 978-981-287-347-7 (eBook)
DOI 10.1007/978-981-287-347-7

Springer Singapore Heidelberg New York Dordrecht London

Softcover reprint of the hardcover 1st edition 2015

Printed on acid-free paper

Springer Science+Business Media Singapore Pte Ltd. is part of Springer Science+Business Media (www.springer.com)

John Vong—*Devon, Neuman, and Susan*
Insu Song—*Yi Pin, Lulu and Cisu*

Foreword

The potential of nations—how successful they may be—is largely determined by just three factors: geography, demographics, and technology. How well the nations achieve their potential is largely dependent on only two factors: economic policy and political culture, for they dictate how talent is gathered and developed to deliver success for the nations.

Technology is a major enabler of success. It can overcome geographical limitations. It can turn the table on destabilizing demographics, whether it be a population with low literacy, or an aging population. Technology allows a "leapfrogging" over the barriers created by geography or anthropologised by demographics. Thus it has particular importance for developing countries, which need not be constrained to follow the same developmental path as today's developed countries, but can instead leapfrog with the emergence of new, more efficient, and more effective technology applications.

The subject of emerging technologies for emerging markets offers critical insights and opens unbounded possibilities on what developing countries can become, and how they can speed-up to get there. To anticipate these emerging technologies is to position a nation well for the unfolding opportunities. This is what makes this reference book for business and technology particularly significant and useful.

The authors have uniquely focused their research on the things hoped for, but are as yet unseen. Their discourse looks at the current and postulates toward the future. The result is a book which is immediately useful for those whose focus is to create the future rather than take incremental steps from the present. It is for those who want to move early and quickly, and to do it not in a haphazard manner but with thoughtful understanding and purposefulness.

The discourse can be read from many aspects. From the education standpoint, the smart technologies discussed follows an education track often referred to as STEM – science, technology, engineering, and mathematics. These are the important subjects to be taught at schools, and to be inspired by teachers, to ensure economic progress in rapidly developing countries. The knowledge of STEM sets the foundation for real goods and services to be produced.

From the technology innovation perspective, the book introduces low-cost ways to deliver the most basic and necessary services to society. This idea rests on frugal innovation, an appropriate terminology for emerging economies, where capital is scarce and risks are high. Yet this is not a book on theories of technology. It dives immediately into applications of technology using frugal approaches to empower communities for a sustainable society. The focus of the technology applications is on education, health, and financial payments, which are the most critical enablers for developing countries to create opportunities for improving the lives of their citizens.

The authors have cleverly woven these different aspects together to call their book Emerging Technologies for Emerging Markets. It makes interesting and important reading for technopreneurs, inventors, angel investors, development specialists, and public policy makers to bring rapid progress to many nations. The policy makers of emerging economies should carefully consider the implicit or explicit recommendations of this book because they offer real insights into making things happen, bring progress, and create a differentiation from countries that are simply expecting or waiting for things to happen.

Finally, I take particular pride in noting that this book is being written, published, and launched in Singapore. Once an emerging economy, Singapore, by embracing technology under a bold and forward-looking leadership, is a living example of what and how a nation can deliver, over a space of just 50 years, on the standards of living and the levels of education, health, and financial services, which much of the world wonders about and hopes for.

Lim Siong Guan
Group President, GIC and Professor in the Practice
Lee Kuan Yew School of Public Policy
National University of Singapore

Preface

Many thought leaders have articulated that, within the next 10 years, the emerging markets will be deeply influenced by emerging technologies such that business operations, societal interactions, and the geopolitical landscape will be significantly changed. These emerging technologies are related to information technology, wireless data communication, man–machine communication, on-demand printing, bio-technologies, and advanced robotics. The emerging markets, regardless of how they may be called—BRIC, BRICS, or BRIICS—will be the main consumers of these new technologies. The markets will, in return, offer ideas to influence technological innovations that are more affordable. By then, firms may have to shift their product development, sourcing and marketing functions to these emerging markets to reap the benefits of affordable innovations. These new technologies and innovations are likely to focus on the three pillars of economic and social developments—financial services, health, and education—due to the multi-cultural settings and demographics of the population in the emerging markets.

There are three trends that challenge the stability of economic and social development in the emerging markets. The first trend is the rising cost of financial services. It will come at a time where only the wealthy are able to have a bank account. The current service charges levied by financial institutions to transact, transfer, and make payments are high enough to push customers away to more cost-effective alternatives. The unbanked populations, which may be as high as 80% of the total national population in emerging markets, living at the base of the pyramid, need access to affordable financial services. The second trend is the rising cost of public and private health services, which has become a deterrent for many to seek medical care. Alternative low-cost approaches for conducting a diagnosis, recommending a prognosis, and providing therapy have to be found. The third and last trend is the huge demand place upon the education systems that will rise with population growth. Students and parents will increasingly face the escalating cost of education.

The authors have dedicated their lives to research on emerging technologies that will drive the sustainable development and growth of emerging markets. The authors strongly believe that these technologies will innovate in tandem with the growth of the emerging markets. Furthermore, the focus of the new technologies will be related to financial services, health, and education. The technologies discussed in this book are affordable, easy to implement, and frugal. For these reasons, the audience for this book is likely to be like-minded people who intend to infuse such technologies and their applications into these rapidly growing markets. Based on the three trends stated above, they are presented in 14 content-filled chapters: Financial Services (Chaps. 2–6), Healthcare (Chaps. 7–10) and Education (Chaps. 11–14).

The authors hope that this book will inspire the creation of new technologies and frugal innovation. Hopefully, it will ignite a creative spark in research and development directed towards emerging technologies in emerging markets.

Acknowledgments

We wish to thank all our collaborators and contributors who provided both overt and covert encouragement, and intentional or unintentional inspiration, to bring this book to completion.

We are grateful to those who have silently encouraged us in word and in action —Mr. Xanana Gusmao (Prime Minister of Timor-Leste), Madam Nguyen Thi Phuong Thao (Vice Chair, Vietjet Air Vietnam), Professor Gunasekaran Anggapa (University of Massachusetts Dartmouth, USA), Mr. Lim Peng Khoon (Former Independent Director, Industrial Bank of China), Mr. Soekardi Hoesodo (Former Deputy Chairman, Board of Financial Development Supervision, Indonesia), Mr. David Hosie, (formerly of HSBC Group), Mr. Dang Van Thanh (Chairman, Thanh Thanh Cong, Vietnam), Professor Purnendu Mandal (Lamar University, USA), Mr. Subinay Nandy (Resident Coordinator of UN Sri Lanka), Mr. Simon Andrews (Country Manager, International Finance Corporation), and Late Professor Tan Teck Meng (Singapore Management University). We thank the World Bank, International Finance Corporation, United Nations Development Program, Asian Development Bank, and the governments of Australia, Indonesia, Vietnam, Timor-Leste, Lao, Cambodia, Myanmar, Philippines, Malaysia, and Singapore. Collectively, they have given project assignments that allowed us to have insights into the economy and social behaviors of, and the impact of technology innovations on, the rural and unbanked populations. These insights are enough to provide fodder for a lifetime of creating technology innovations for the emerging markets.

We are grateful to all who have reviewed, edited, proofread, and formatted this book within a tight deadline. They had to focus on the key messages that emerged from the various papers. We particularly thank participations, contributions, and feedback from Shah Atiqur Rahman, Baying Lei (Nanyang Technology University, Singapore,) and postgraduate students at James Cook University Singapore Campus who reviewed the first drafts. The book could not have been published without the help of our able research assistant Ms. Kar Ling Ho. She was a blessing to us in proofreading each chapter and created charts and tables. We also wish to thank Loyola D'Silva and Ravi Vengadachalam from Springer. Loyola started us on this writing journey while Ravi kept pushing us to completion. Any deficiencies that remain would now be the sole responsibility of the authors. Finally, there are still many others who assisted along the way. You know who you are.

John Vong
Insu Song

Contents

Chapter 1
Introduction

Abstract The modern digital revolution that encompasses the computers, connectivity, data, iPhones and internet, is both disruptive and divisive. It is disruptive because it can and has displaced millions of workers. It is divisive because it divides those who are experienced and technology-savvy from those who are just experienced. The former will have promising career options and business opportunities while the latter will see their experience and their jobs replaced by intelligent technology. This disruption and division will significantly impact the emerging markets. Take for example, Foxconn, an icon of Chinese manufacturing, at its height of its production employed 1.5 million workers to assemble electronics. With the rising cost of labour and falling cost of automated manufacturing, it has since replaced workers with robots. China's path of progress will see more companies similar to Alibaba that recently made a spectacular debut on the New York Stock Exchange that employs only about 20,000 people.

Keywords Emerging technology · Emerging market · Innovation · Financial service · Health service · Education

1.1 Technologies that will Reshape Our Future

In general the emerging technologies encompass new technologies that are currently developing, or will be developed, over the next 5–10 years, and which can significantly change the business models and social environment. These include information technology, wireless data communication, man-machine communication, on-demand printing, bio-technologies, and advanced robotics.

The World Economic Forum's Global Agenda Council on Emerging Technologies reveals recent trends in technological changes in its Top 10 Emerging Technologies. By identifying the most important technological breakthroughs, the Council aims to raise awareness of the potential and to guide investment, regulation and public understanding of such emerging technologies. For 2014, the Council

J. Vong and I. Song, *Emerging Technologies for Emerging Markets*,
Topics in Intelligent Engineering and Informatics 11,
DOI 10.1007/978-981-287-347-7_1

identified 10 new technologies that could reshape our society in the future: Body-adapted Wearable Electronics, Nanostructured Carbon Composites, Mining Metals from Desalination Brine, Grid-scale Electricity Storage, Nanowire Lithium-ion Batteries, Screenless Display, Human Microbiome Therapeutics, RNA-based Therapeutics, Quantified Self (Predictive Analytics), and Brain-computer Interfaces (World Economic Forum 2014).

1.2 The Emerging Markets that will Define the World

The term emerging market first appeared in 1981 and was reported by economists at the International Finance Corporation (IFC) when the group promoted the first mutual fund investments in developing countries. Antoine van Agtmael who coined the term, emerging markets, now runs a fund management company with US $24 billion invested in those markets. He believes that the combined GNP of today's emerging markets will overtake that of the developed economies. In the next 50 years, emerging markets will be twice the size of today's economic leaders. Assuming that information technology is deployed to full effect, the populations of these countries would be wealthier, healthier and better educated (The Economist 2007).

Since then, references to emerging markets have been popularized by the media, used in foreign policy and trade debates, and proliferated in investment fund prospectuses and multinationals' annual reports. The definition of the term varies widely to the referenced countries and depending who is reporting it (Khanna and Palepu 2010).

Two decades later, a former Goldman Sachs economist came up with the term "BRIC" that comprises of Brazil, Russia, India and China as countries whose growth will shape the world economy in the coming decades (Cocks 2014). Later "S" was added to become BRICS that included South Africa. There is also another version BRIICS which included Indonesia and South Africa.

The Asian Development Bank (ADB) has identified 25 countries as emerging market economies: Argentina, Brazil, Chile, China, Columbia, Czech Republic, Egypt, Hong Kong, Hungary, India, Indonesia, Israel, Rep of Korea, Malaysia, Mexico, Peru, Philippines, Poland, Romania, Russian Federation, South Africa, Singapore, Taipei, Thailand, and Turkey. Of the 25 economies, 11 are in Asia, seven in Europe, five in Latin America, 1 in the Middle East, and 1 in Africa (Park and Mercado 2013).

In mid-2013, Bloomberg listed the top 20 emerging markets as China, South Korea, Thailand, Peru, Czech Republic, Malaysia, Turkey, Chile, Russia, Indonesia, Columbia, Poland, Namibia, Zambia, South Africa, Mexico, Brazil, Hungary, Morocco and Philippines. The list of markets is not that different from ADB, with the exception it is more inclined towards the African continent (Bloomberg Markets 2013).

However The Economist (2013) reported that the most impressive growth was in four of the biggest emerging economies: Brazil, Russia, India and China (BRIC). These economies have taken their own growth paths and for different reasons. But their size singled them out as special, in terms of purchasing-power, they were the only $1 trillion economies outside the OECD. Their growth rates were spectacular (see chart). Today they are four of the largest ten national economies in the world.

In 2010, a popular US political commentator said that the "rise of the rest" is a transformative, tectonic shift in the distribution of global power (Khanna and Palepu 2010). He reported that companies based in these emerging economies are already challenging multinationals based in the developed world. Take for examples, China-based Lenovo has purchased the IBM personal computer business in 2004, and the acquisition of Jaguar and Land Rover by Tata Motors of India in 2008 are signs of the increasing business expansion by firms in emerging markets. Some observers see the financial crisis of 2008 as an inflection point which accelerated the emergence of these markets as dominant players in the global economy.

1.3 R&D, Innovation and Technologies in Emerging Markets

The OECD Report 2014 revealed that OECD countries on average invested 2.4 % of their GDP in R&D in 2011. In the emerging markets, China invested 1.8 % of its GDP (second highest spender on R&D in the world for 2011), followed by Brazil 1.2 % (2010), Russian Federation (1.1 %), South Africa 0.9 % (2008), India 0.8 % (2009), and Indonesia 0.1 % (2009). Although they invested less than OECD countries, China has made great progress in the last decades (OECD 2014, p. 80) (OECD 2014). The report further stated that productivity among firms in China, Indonesia, and South Africa was higher than those firms that did not have R&D spending (OECD 2014).

The rapid growth in interest and importance of technology innovations in emerging markets is evidenced by the exponential increase of published articles about innovation (Jana 2009; Saraf 2009).

Although there is much R&D investment in the emerging markets, there is a distinct approach being undertaken. The brand of innovation is called "jugaad" in India, "zizhu chuangxin" in China, "gambiarra" in Brazil and "jua kali" in Kenya. The English translation is associated with DIY or "make do and mend or just do it (Radjou et al. 2012)."

The multinational corporations led the world in R&D in new product development. For the vast majority of consumers in emerging markets, they consider these products as expensive, not user friendly and with many features not deemed necessary. There is a need to keep the product simple that directly bring immediate benefits.

There is the principle of reverse innovation that says that firms will need to start moving product development, sourcing and marketing functions overseas (Govindarajan et al. 2012). The mismatch in applications and pricing between developed and emerging markets means that the dependence on Western product development is not easily sustainable.

The farmers of South Africa's Limpopo province use mobile phones to access new markets to sell their (agriculture) produce provides evidence that the technology divide can be narrowed (The Economist 2007). In the same breath, the use of a village phone can help the low-income households in rural areas of Nigeria with employment opportunities, and thus to make a living and improve their quality of life (Joseph 2012).

In recent years, information and communication technologies (ICT) impacted greatly socioeconomic development in developing countries. Encouraged by the development of rural IT network projects in countries like India, Sri Lanka, Bangladesh and Philippines, the government of Vietnam focused on IT infrastructure and capacity building efforts. Under the framework of public administration reform in Ministry of Agriculture and Rural Development (MARD), a study was conducted in setting up Rural Telecentres (RTs) to provide information to farmers in Vietnam. Farmers' opinions and their needs in relation to agricultural and rural issues were considered in setting up 13 RTs in rural communities. The post-implementation survey indicated that the RTs fulfilled their objectives in improving the livelihoods of rural farmers (Vong et al. 2013).

The authors believe that the technologies that will emerge in tandem with the growth of the emerging economies will be in health, financial services and education.

1.4 Health Technologies for All

The WHO (2011) has published a compendium of new and emerging technologies that addresses global health concerns. These technologies are likely to be suitable for use in low-income environments. The publication aims to increase dialogue between ministries of health, procurement officers, donors, technology developers, manufacturers, clinicians, academics and the general public. In doing so, WHO hopes to raise awareness of the importance for appropriate design solutions, and for further development and technology dissemination (WHO 2011). The compendium specifically focuses on innovative technologies that are not yet widely available in developing countries and some product concepts are still in progress. Technologies in the compendium are presented in one page that summarises the health problem addressed, the proposed solution and product specifications, based on data and information provided by the developers of the technologies concerned.

The eligibility for inclusion in the compendium has been evaluated by EuroScan member agencies and WHO. Forbes (2014) reported that the list of projects among many others are as follows: assisted vaginal delivery instrument; fetal heart rate monitor by mobile phone, infant warmer, isolator system for laparoscopic surgery, lab-in-a-backpack, point of care screening/diagnostic, low-technology child restraint car seat, microbial water testing kit, mobile health record system for pediatric HIV, mobile phone image transmission for diagnosis, mobile phone pulse oximeter, pedograph, point-of-use water purifier, portable cell sorting and counting device, portable system for pre-cancer screening at point of care, portable telemedicine unit, portable transcutaneous haemoglobin meter, and single-size contraceptive diaphragm (Honigman 2014).

In emerging markets with the vast populations, mobility is the key to health services. Firstly, mobile-focused healthcare advances are simplifying the way new technology is adopted by medical professionals and organizations. Secondly, the wearable technology, including Misfit Wearables, FitBit, Jawbone UP and Nike Fuel Band, will continue to provide medical professionals with the data they need to effectively monitor and treat their patients on a day-to-day basis. Lastly, the EHR Software electronic health records (EHR) software is providing governance and accountability, control and the opportunity for healthcare providers and their patients.

The Economist (2007) has reported a health technology project led by Afghanistan's biggest telecommunications company and partnering with the French Medical Institute of Children (FMIC) and the Aga Khan University Hospital (AKUH) in Pakistan, where FMIC will provide teleradiology (electronic transmission of radiological patient images) (The Economist 2007). There will be an average of 60–80 transmissions and 10–15 teleconferences between hospitals per month, with the numbers increasing over time.

Other examples of health technology innovation for emerging markets are breathing diagnosis (Lei et al. 2014) and programming a mobile phone application to perform diagnosis of Bell Palsy using the House Brackmann score (Diederich 2013).

1.5 Financial Services Technologies for the Under-Banked

Most researchers agree that growth in the banking sector in terms of customers, deposits and lending is likely to come from the emerging markets. In these markets, about half the world's adult population do not have a bank account at a formal financial institution, whether it is a bank, a savings and loan association, or a credit cooperative, according to a recent household surveys undertaken by the World Bank and Gallup.

Although the emerging markets have different profiles in terms of depth of banking, many have low levels of bank usage. This offers the significant potential to engage new customers and deepen the financial sector. For example, in China only some 64 % of adults have a bank account, with lower levels in Brazil (56 %), Russia (48 %), India (35 %) and Vietnam (20 %).

By contrast, in developed markets most adults have bank accounts. Any growth in customer numbers depends on population growth and immigration, which are often stagnant. For example, in the United States 88 % of adults hold bank accounts, with even higher rates of bank usage in Japan (96 %), the United Kingdom (97 %) and Germany (98 %). The industry has recently been shrinking in many rich countries as banks trim loan books, sell off assets and realign their capital ratios to meet regulatory requirements.

New technologies are important drivers in the rise of financial institutions in emerging markets. Of late, mobile banking and mobile payments have received the most attention, particularly in Africa, where few people have bank accounts but many own mobile phones. There are emerging technologies that support many other technology applications, such as wireless Internet access. It allows banks to reconcile financial transactions as they take deposits, honour withdrawals and grant loans from any location. Meanwhile, new payment systems increasingly allow for instantaneous digital transactions without cash, a bank branch or an automated teller machine.

Safaricom already has more than 450 M-Pesa agents across Kenya, compared with just 350 for Western Union. Launched in March 2007, M-pesa money transfer service attracted about 150,000 users by June. An estimated 2,500 new users were signing up daily, such that in the first three months, the value of transfers amounted to nearly US$6 million. The average transaction size of US$45 appears high. This is possible because M-Pesa is also being used for commercial transactions such as paying wages to workers, besides the small personal transactions.

Safaricom, partly owned by the UK's Vodafone, expects more than 1 million people to be using the M-pesa service by 2015.

India's largest mobile operator, Bharti Airtel, launched a nationwide mobile-wallet service—Airtel Money—in February 2012. The Airtel Money is the result of technology collaboration between a bank and mobile operator which aims to provide access to financial services for the vastly dispersed communities in India. Bharti Airtel service allows users to top-up cash onto their mobile devices. The users can then use the cash top-ups to pay utility bills, make mobile recharges, shop at retail outlets and transact online. It is the first nationwide mobile-based service in India to offer on-line money transfers from one Airtel Money wallet to another, or to bank accounts (The Economist 2012).

The emerging technologies for banking in emerging economies are likely to focus on payments transfer between banks and customer, between individuals to individuals, and between government agencies and citizens. The delivery of financial services though mobile phone devices in Cambodia (Vong et al. 2012) and advent of cashless banking (Song and Vong 2013) are dawning.

1.6 Education Technologies for the Masses

There are possibly five trends that warrant our attention that will definitely affect education, especially in emerging markets because of the size of their young population. Firstly, there is the Cloud Computing. Virtual learning saves time and resources. More money and more teachers are needed for the traditional brick-and-mortar education model. For countries that have scarce financial and manpower resources, virtual learning is a good approach. Secondly, is the use of mobile technology. These mobile devices are more affordable than laptops and require little infrastructure. Thirdly, Game Based Learning as gamification has become an effective teaching and learning tool. Fourthly, it is MOOCs, Massively Open Online Courses. It allows students to learn at their own time, own pace, and free of charge.

Finally, it is Learning Analytics that will help teachers to analyse their data to observe students learning behavior, trends, and patterns to improve their teaching styles.

The youthful population of Nigeria, Myanmar, Mexico, Indonesia, and Turkey needs much education. The education infrastructure in terms of school and teachers presents a challenge to the traditional bricks and mortar approach to learning. New learning paradigms are now being researched and experimented in emerging economies. A case in point is automated tutoring system, where students can learn anywhere they like, at home or in the field or while shepherding a gang of goats, with the use of a tablet. The curriculum is embedded into the tablet at start of the school term. The students will go through the self-paced learning and self-directed assessments with multiple choice questions. At the end of the semester, the student then takes a final exam at the school. A possible experiment is being carried out based on automated tutoring systems (Song and Vong 2013).

The Vietnamese government is seeking to develop an information society, digitalizing communications and making electronic conferencing a mainstay of organizational life. (Vietnam Institute for Education Sciences and VVOB. 2009.) To meet these goals, the Vietnamese government has promised 8.4 billion USD in ICT investment between 2010 and 2022, as well as favourable licensing and business support for companies developing Viet Nam's IT capabilities (UNESCO Asia Pacific Regional Bureau for Education 2013).

In Thailand, the level of ICT infrastructure is stretched to support the implementation of advanced ICT in Education projects for the public school system. In their Second National ICT Master Plan (2009–2013) it admitted that the main factor holding back the development is the readiness of the information and communication infrastructure (Ministry of Information and Communication Technology 2009).

References

Bloomberg Markets (2013) The top 20 emerging markets. http://www.bloomberg.com/slideshow/2013-01-30/the-top-20-emerging-markets.html. Accessed 06 Oct 2014
Cocks T (2014) Jim O' Neill: BRICs, MINTs strong despite emerging market wobbles. Reuters. Accessed 06 Oct 2014
Diederich ISNYYJVJ (2013) Profiling bell's palsy based on house brackman score. J Artif Intelligience Soft Comput Res 3(1):41–50
Govindarajan V, Trimble C, Nooyi IK (2012) Reverse innovation: create far from home, win everywhere. Harvard business review press
Honigman B (2014) Three emerging tech tools affecting healthcare. Forbes
Jana R (2009) India's next global export: innovation. Bloomberg businessweek. http://www.businessweek.com/innovate/content/dec2009/id20091221_864965.htm. Accessed 13 July 2009
Joseph MK (2012) The african experience with ICT for rural women's development. Rural development—contemporary issues and practices, InTech
Khanna T, Palepu KG (2010) How to define emerging markets. Forbes. Accessed 06 Oct 2014
Lei B, Rahman SA, Song I (2014) Content-based classification of breath sound with enhanced features. Neurocomputing 141(0):139–147. http://dx.doi.org/10.1016/j.neucom.2014.04.002
Ministry of information and communication technology (2009) The second Thailand information and communication technology. Master plan (2009–2013)
OECD (2014) Society at a Glance 2014: OECD Social Indicators.
Park C-Y, Mercado RV (2013) Determinants of financial stress in emerging market economics. Economics working papers
Radjou N, Prabhu J, Ahuja S, Roberts K (2012) Jugaad innovation: think frugal, be flexible, generate breakthrough growth. Jossey-Bass
Saraf D (2009) India's indigeneous genius: jugaard. Wall Street J
Song I, Vong J (2013) Mobile core-banking server: cashless, branch and wireless retail banking for the mass market. In: International conference on IT convergence and security (ICITCS), pp 1–4. doi:10.1109/ICITCS.2013.6717857
The Economist (2007) Path to growth: the power of technology in emerging markets. The Economist
The Economist (2012) Beyond branches banking report. The Economist
The Economist (2013) Emerging economies: when giants slow down. The Economist. Accessed 07 Oct 2014
UNESCO Asia Pacific Regional Bureau for Education (2013) ICT in education policy, infrastructure and ODA status in selected ASEAN countries.
Vong J, Fang J, Song I (2012) Delivering financial services through mobile phone technology: a pilot study on impact of mobile money service on micro-entrepreneurs in rural cambodia. Int J Inf Syst Change
Vong J, Song I, Mandal P (2013) Application of ICT to improve rural livelihood in Vietnam. Int J Electron Finance 7(2):132–145. doi:10.1504/IJEF.2013.057278
World Economic Forum (2014) Top 10 emerging technologies 2014: from the global agenda council on emerging technologies. World Economic Forum
World Health Organization (2011) Compendium of new and emerging health technologies. World Health Organization

Part I
Financial Service

Chapter 2
Mobility Technology Solutions Can Reduce Interest Rates of Microfinance Loans

Abstract In recent years there is intense criticism of the exploitative interest rates charged to borrowers of microfinance institutions (MFIs). Development specialists and researchers have found incidences of usurious lending rates being levied upon the unbanked population, ironically in an attempt to lift them out of poverty. The borrowers suffer greater indebtedness and take extreme measures to repay while others just defaulted. One important reason for the high lending interest rates is the transaction cost or sometimes called administrative cost or operating expenses. It presupposes that high lending interest rates are necessary to cover the transaction and administrative expenses (TAE) of providing many small loans and monitoring many small borrowers. Thus this paper aims to identify the cost drivers and the cost of the MFI transactions and administration by using activity based costing (ABC). Thereafter a mobility solution is designed to replace the manual processes to lower the TAE, which in turn reduces lending rates charged to MFI borrowers. A pilot project has been started in Indonesia to test the hypothesis.

Keywords MFIs · Microfinance · Microcredit · ABC · Costing · Interest burden · Systems technology

2.1 Introduction

The principal function of banks is to provide their banking customers with access to reliable and affordable financial services such as savings, credit, payments, transfers, and insurance. These are important services that help any modern society and a healthy economy. However half of the world population is still unbanked.

In the last few decades, many MFIs have emerged to provide financial services and microcredit to the unbanked sector. For example, solidarity lending and village banking pioneered by Grameen Bank in Bangladesh (Yunus and Jolis 1999; Hughes and Lonie 2007) and KWFT (Microfinance) in Kenya, WING in Cambodia (WingMoney), and many others (KWFT 2014; Vong et al. 2012). It is believed that

J. Vong and I. Song, *Emerging Technologies for Emerging Markets*,
Topics in Intelligent Engineering and Informatics 11,
DOI 10.1007/978-981-287-347-7_2

microfinance is an effective instrument to alleviate poverty by providing financial service, particularly microcredit and savings accounts to the low-income individuals and microenterprises (Hammond and Prahalad 2002).

However research conducted by Chaia et al. (2009) and Ardic et al. (2011), suggests that as at end 2009, there are 2.75 billion people (56 % of the global population) who do not have access to formal financial services. A recent World Bank survey of 150,000 adults aged 15 and above in 148 different economies reported that over 50 % of the respondents do not bank with a financial institution. While 89 % of adults in developed countries have access to financial services, it is a paltry 41 % in developing countries.

The unbanked group, usually individuals living under US$5 per day, is caught in a dilemma. Firstly, most banks do not want them as customers because they are not bankable, that is the bank cannot possibly see how they can make a profit from this group. A bank will need to cover the cost of funds (that is, cost of micro deposits or cost to obtain the capital), cost of loan default (when borrowers do not repay), and the cost of administration (staff cost, systems, premises and utilities).

Secondly, the unbanked may not pass the normal know-your-customer (KYC) process because of lack of identification papers and a permanent residence. Thirdly, the unbanked group will find the cost of financial services exorbitant. Imagine, this unbanked person walks into a bank to request for a money transfer of $20 to a relative in another city, and the bank charges a transfer fee of $5—a significant 25 % is deducted from the transfer.

Lastly, the distance of the nearest branch or ATM is far away from the place of residence such that the unbanked person takes a long time to access the financial services. Thus one-third of the unbanked population is in Asia, approximately 800 million people, who are in the lowest income category (that is, living on under $5 per day), simply cannot afford the high costs of banking services. From the banks' point of view, it is a challenge to provide a self-sustainable business model to serve these markets.

2.2 Why Is MFI Interest Rates So High?

From the analysis of data of several countries from 2004 to 2011, the researchers have concluded that operating cost is the largest determinant of interest rate levels in microcredit institutions (Rosenberg et al. 2013).

There are huge challenges in delivering loans to low income households in developing countries. Banks and other financial institutions have failed to meet the financing needs of the poor in developing countries. The reasons for the neglect of this large group lie in the nature of traditional banking practices which are based on guarantees, reputation, and collateral. It is good banking practice for bankers to have an intimate knowledge of market conditions, which helps them to assess the initial loan applications and then continuously monitor the borrowers (Koveos and Randhawa 2004).

Collateral helps align borrowers' intentions with those of the bank and provides a security in case of loan defaults. Banks also need timely and accurate information and continued updating of accounting records. However the low-income households and individuals are the primary clients of microfinance institutions. They do not have financial resources of financial assets to offer as collateral to guarantee their loans. They do not normally have a credit history. As such they are excluded from the formal financial sector. Banks would regard these groups as having a high credit risk profile (Robinson 2001).

The case of adverse selection arises when banks, lacking credible information about the credit risk of borrowers, cannot differentiate between borrowers with varying degrees of credit risks, and resulting in charging high interest rates on all borrowers. Higher interest rates in turn drive creditworthy borrowers out of the credit market, leading to an inefficient outcome (Aghion and Morduch 2004).

Majority of existing microfinance solutions are operated on labour-intensive work processes that involve manually writing or transcribing transactions from pieces of papers to ledger cards. These will present both a challenge and risks the MFIs. They will find difficulty to increase their capacity to serve the huge demand of the unbanked population. The risks that they face are genuine mistakes in manually writing and transcribing accounting records and that of frauds. For certain, a substantial portion of transaction and administrative expenses (TAE) is attributed to manual labour.

MFIs that provided loans to low-income borrowers in developing and transition economies have charged high lending rates to cover their costs (Rosenberg et al. 2009). Most MFIs believe that this is the most appropriate approach to ensure financial sustainability and further growth. MFIs that are profitable may continue to serve their borrowers without ongoing donor contributions and may even fund growth by drawing on commercial sources of funds, including deposits from the public.

There is much evidence that the focus on financial profitability and efficiency have increased the borrowing cost to the poor. Moreover lending money to the low-income borrowers or the rural poor can be costly. MFI Agents have to maintain frequent contact and close relationships with customers to enforce loan contracts and to control lending risk. Thus the twin goals of social outreach and profitability are in conflict (Hermes and Lensink 2011). Their ability to provide collateral for their borrowing is also not easily available.

In a CGAP study, the lending rate of MFIs range from 17 % per annum (in Sri Lanka) to 80 % per annum (in Uzbekistan), with a global average of 35 % per annum (CGAP 2008). The high microcredit interest rates have been heavily criticized and a number of countries have tried to regulate an interest rate cap on microcredit lending (BBC 2013).

2.3 Reducing Operational Cost

Existing literature indicates that transaction costs are a major contributor to high interest rates on microcredit loans (Shankar 2007).

The limited resources and the sheer numbers of borrowers present a tremendous hurdle to process individual loan applications. To create a sustainable self-funding model for a growing MFI, is to design a system that lowers monitoring costs, create a committed and creditworthy client base, and help poverty alleviation (Koveos and Randhawa 2004).

Few studies have examined measures of MFI efficiency. The accounting variables to assess the efficiency of the MFIs are—administrative expense ratio, number of loans per loan officer, loan officers to total staff, portfolio size, loan size, lending methodology, source of funds, salary structure.

In Lafourcade et al. (2006)'s study that uses cost per borrower and cost per saver as measure of efficiency, it was found that African MFIs incur highest costs per borrower but lowest costs per saver. They also discovered that regulated MFIs maintain higher efficiency through low costs per borrower and per saver. In contrast, African cooperative-MFIs are the least efficient with the highest cost per borrower. Nevertheless, cooperative MFIs have the lowest cost per saver but unregulated MFIs have the highest. None of these studies use any parametric or non-parametric approach to evaluate the efficiency of MFI.

Intuitively, the approach is to identify the processes that increase TAE. Only then can the process be changed or reduced or to fully be replaced by low-cost processes.

For example, it is noted that it is costly to monitor borrowers. So microfinance experts have suggested social collateral. This approach works through reputational effects on their community in which repayment of loans is seen by group member. The borrowers are to pay their loans in a "ceremony" for all to see. In this way, the borrowers are able to maintain their social standing in the community (Woolcock 2001). Goldmark (2001) suggested other methods that may help build social collateral, thereby making loans even more secure. Van Tassel (1999) proposed a model with a one-period game to determine the optimal group-lending contract under asymmetric information. He concludes that agents will always form groups with agents of the same type. The agents' types can be differentiated by the level at which they are willing to trade-off increased joint liability commitments for lower interest rates.

Ghatak (1999) concluded that group lending not only increases repayment rates and welfare via social collateral, but also due to peer selection by members of the lending group. Many Indian MFIs, for example, reduce their staffing costs by lending to self-help groups rather than to the individual borrower. Yet another study show that bank-MFIs are the most efficient under intermediation approach while NGO-MFIs are the most efficient under production approach (Haq et al. 2010). Similar to Ghatak, the study by Islam (1995) concludes that lenders using peer-monitoring systems can charge lower rates relative to conventional lenders. Furthermore, at the same interest rate, the expected rate of repayment is higher with lower risk (Islam 1995).

2.4 Monitoring Transaction Cost and Administrative Expenses Using Activity Based Costing (ABC)

Activity-based costing (ABC) was first propounded in 1987, but the dearth of the available data published (Gosselin 2006), suggested that ABC is not as widely understood and practiced in the financial service industry as one might be led to believe (Wegmann 2011; Rasiah 2011; Lee 2002). Two reasons account for this: firstly, financial institutions find it difficult to define their products (Schoute 2011) and secondly, banks think that the difficulty involved may not warrant the cost (Askarany and Yazdifar 2007; James 2013). In Asia, the first study of applying ABC to banking was undertaken by Vong (1994).

The concept of ABC presents a basis for understanding the behavioural patterns of all cost items within an organization, linking operating costs in a value chain in such a way that management can identify the factors that drive expenditure, focus on key cost items and manage them more effectively. ABC enables management to monitor the factors that drive costs and control the implications of organization's profitability (Hicks et al. 2009).

The primary function of ABC is to assist management in decision-making (Kee and Schmidt 2000). It is an approach in cost analysis that helps management to analyse its cost base in a more meaningful way. It analyses cost behaviour by activities, linking costs to factors that are responsible for the occurrence of those expenditures, and thus, focusing on the factors more effectively in order to manage costs (Bogdănoiu 2009).

As shown in Fig. 2.1, ABC draws information on the activities of all departments within an organization. It depicts a clearer picture of all inter-relationships among the activities and costs incurred. Management uses ABC to understand how and why certain costs were expended, thereby helping them to focus on the root problems, in the business (Qian and Ben-Arieh 2008).

Fig. 2.1 ABC model

Estrin et al. (1994) suggested that the basic premise of ABC was period cost. For example, indirect, variable and fixed costs were attributable to the activities required in making various products. The authors further reported that many non-manufacturing costs, such as marketing costs, were also a function of product activity and should be loaded into products responsible for their occurrence.

An early example of ABC applications in the banking and finance is the First Tennessee National Corporation. By its own admission, it had tried a sophisticated costing system developed for banks by a leading public accounting firm but it did not perform as required.

The corporation switched to ABC in 1989, which is the focus on certificate-of-deposit (CD) products. The costing system revealed that 30 % of the firm's customers accounted for 88 % of its profit. The system also revealed that the firm was losing money on the last 30 % of its customers. An analysis revealed that the loss was due to the fact that one could purchase a 90-day $500 CD. Given a one percent spread, it was estimated that the bank generated revenue of only five dollars a year if they had reopened a $500 CD four times a year, where the revenue obtained was considerably less than the processing cost of the transactions. By simply raising the minimum CD amount, the bank eliminated the loss, thus increasing the overall profitability of the CD product line.

Kantor and Maital (1999) built a method for measuring product-specific inefficiency in bank branches to facilitate precise measurement of waste and identify its causes. Their approach combines and integrates two separate, widely used models for measuring costs and efficiency performance: activity-based cost accounting (ABC) and data envelopment analysis (DEA) (Kantor and Maital 1999). The resulting method for activity-based management provides managers with detailed quantitative performance benchmarks for the specific business activities of their firms or divisions. One of the key issues in banks is that lending managers need to assess the profitability of a customer relationship (van Raaij et al. 2003). This would mean deducting the cost of servicing the account from the revenue obtained from that customer. Since banks provide a wide range of products to clients, especially in corporate banking, lending managers need to know the type and volume of products transacted and more importantly, the transaction costs of those products (Chea 2011). This would enable them to find an appropriate mix of banking facilities offered to the client, which would enable the bank to achieve a targeted level of profitability. This would mean that the transaction cost of each product type must be calculated (Nenovski et al. 2012). Three hurdles need to be overcome before that can be achieved: the products need to be defined, the cost structure agreed on and the process drivers determined (Heisinger and Hoyle 2014).

In the study by Vong (1994), he outlined a 4-step approach to build a simple ABC system to monitor costs.

Step 1: Define the Financial Products

Product costing can be regarded as basic form of activity-based costing, therefore forming the basis for product pricing and product profitability (Lee 2002). Kotler (1976) defined product as anything that could be offered for attention, acquisition or consumption. This definition included physical objects, service, personalities, place, organizations and ideas. Hempel (1976) regarded bank products as customer satisfaction. Wills and Day (1984) suggested that bank products should be seen in the light of benefits derived, and that service products were those which produced a series of benefits which could not be stored. However, a bank product is not only a service offered to the customer, but also a measure of creating actual or potential income (Dietrich and Wanzenried 2014). Then there are some financial services, which may not attract income by themselves, but rather contribute towards the main product which does get (Colgate and Alexander 2001). Vong (1994) described them as "a series of benefits" and called them sub-products. He further defined a product as "a service offered directly to the customer that creates actual, notional or potential income".

In our case, most MFIs have few products, namely microloans, deposits and insurance services. In order not to be fazed by the inconsistencies in numbers, the MFI should first survey the broad definitions of products that can be readily applied. Given the agreed definitions, bank products can then be identified and listed.

For example, a cash withdrawal transaction from a Loan Account is a cost and does not provide an income itself but the main product Loan Account attracts notional interest income. Transactions linked to a main product can be referred to as "sub-products" and sometimes defined as "the lowest level of a product where costs can be identified". Thus, a main product may have a series of sub-products which incur cost but which contribute indirectly to the production of income at the main product level.

Table 2.1 shows the possible sub-products contained with the Current Account product.

Step 2: Decide on the Cost Structure of a Product

It is envisaged that the total costs incurred by the MFI can be devolved to all the bank products. This process is split into four three cost channels similar to the principles of Horngren and Foster (1987). Then each product costs have three components: direct cost, indirect cost and overheads.

Table 2.1 Loan product and sub-products

Product name	Product description
Main product: loan service	Loan account services offered to borrower
Sub-product 1: credit evaluation	Evaluating the credit proposal
Sub-product 2: borrower training	Training the borrower prior to loan release
Sub-product 3: loan cash withdrawal	Taking cash out from loan account
Sub-product 4: loan monitoring	Visiting the borrower
Sub-product 5: loan cash Repayment	Collecting the cash for repayment
Sub-product 6: loan account open	Process of opening a loan account

- Direct Costs

Direct costs are those that can be identified specifically with a single cost objective in an economically feasible manner. In an MFI, they would be those operational expenses incurred in processing products. For example, the Loans Department would have its entire administrative expenses routed to the loan transactions—account opening, loan cash withdrawal, and loan repayment. The cost allocated will be called direct costs.

- Indirect Costs

Indirect costs are those administrative expenses, which cannot be identified specifically to the actual loan traction processing but closely associated with it. For example, the Cash Section of the Loans Department. The cost of Cash Section will have their administrative expenses allocated to the Loan transactions as indirect cost.

- Overheads

Costs incurred by departments in providing supervisory/general services to other sections or departments. An example would be the cost of the CEO Office in MFI. This cost will be allocated to the Loan products as overheads.

Step 3: Determine the Process Drivers

ABC is linked to performance measurement (Dickinson and Lere 2003). Companies that have high costs associated with critical activities would want to identify process drivers for those activities and encourage employees to improve those processes. Kaplan and Anderson (2004) of Harvard Business School defined process drivers as factors that affect the efficiency of the activity.

ABC can be viewed as a two-step allocation process. The first allocation, using first-stage driver statistics as basis, distributes expenses to activity pools. The second allocation, using second-stage driver statistics, distributes activity costs to products (Khataie et al. 2011).

While many costing systems assume that the responsibility for costs falls neatly within organizational boundaries, process drivers recognize that this is not the case and instead, focus on decisions which give rise to cost being incurred to volume, quality, time and the level of service.

Process drivers should be easily measured and determined. This gives objectivity in the estimation of different types of cost structures (Homburg 2004). Table 2.2 shows the process drivers that can be adopted by a bank.

Time-based weighted transaction counts take into consideration the processing time for different transactions. The cost allocation is not only on the basis of the number of transactions but also the time taken to carry out those transactions.

In the MFI industry, productive staff count can be used as a basis to allocate indirect costs and overheads. Productive staff count is a summation of the number of direct staff (labour input that can be directly identified with the transaction) and indirect staff (labour input that can only be indirectly associated with transactions).

Table 2.2 Process drivers and cost assignment

Process drivers	Operating expenses to be assigned	Cost type
Time-weighted transaction count	Costs incurred at sections/departments which can be identified specifically with the cost objective	Direct cost
Productive staff count	Cost incurred at sections/departments which cannot be identified specifically with the cost objective but which are closely related to the processing activity	Indirect cost
Productive staff count	Overheads incurred by sections/departments which are distantly related to the processing activity	Overheads

The transactions that require a greater proportion of productive staff would have higher indirect costs and overheads allocated to them.

Step 4: Planned Implementation Phases

One of the most important steps is the implementation of the ABC concept without which all investments in time, manpower and money would be futile. It is necessary to decide on how sophisticated a system is needed to report the required information and how a realistic time frame can be put in place (Liu and Pan 2007). Depending on the nature of the MFI's operations and the complexity of the reporting details, it may take between 6 and 12 months for the development, design and implementation of an ABC system for input to customer and product profitability analysis (Sena 2010).

The implementation activities include product definition and coding, product mapping (linking cost centres and general ledger items to products) and the development, design and testing of cost assignment programmes (Roztocki 2005).

Software solutions like SAP, PeopleSoft and Oracle have ABC components that help to assign overhead costs to customers and products. Push and pull approaches shown in Fig. 2.2 are used for assigning cost centre resources to business processes (via resource drivers) and allocating business processes to receiver (via process drivers).

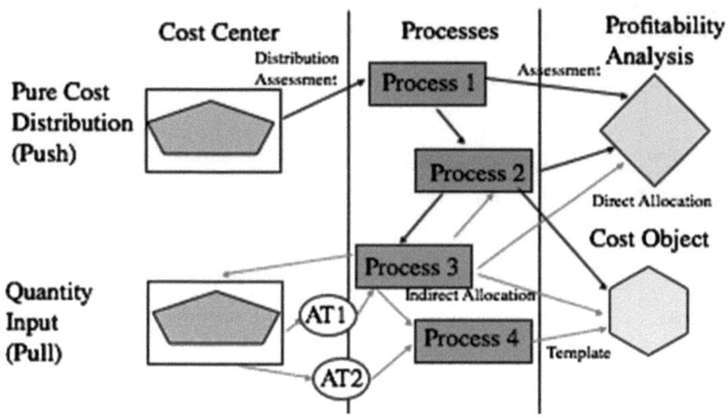

Fig. 2.2 Push and pull factors depiction

2.5 Pilot Project in Indonesia

After identifying the process cost and cost drivers, we may use mobility technology solutions to perform the processes that drive the operational costs. A pilot project has recently being designed and undertaken in Indonesia by the authors. In the pilot project, the entire process from borrower acquisition, loan monitoring, loan repayment, and even peer-to-peer payment is monitored and then to design mobility solution to replace time consuming and laborious processes that increase cost. Figure 2.3 illustrates the overall process.

2.5.1 *Borrower/s Acquisition Process*

The MFI agent takes a long journey from the city to a rural village to visit a potential borrower (or borrowers in the case of group lending). An interview is conducted, with a series of questions, to determine the loan amount and loan repayment ability. The agent returns to the MFI Office at the fringe of the city with the data collected from the potential borrower and of the potential borrower. The MFI Agent analyses the data. If the information from the analysis matches with the lending criteria of MFI, a loan proposal is written including the loan repayment schedule for submission to the MFI loan committee for approval. Not every loan proposal is approved. The usual duration of this process ranges from 5 to 15 days.

2.5.2 *Training of Borrower/s*

After the loan is approved, the MFI agent or another team makes the journey to village for the second time. The purpose of the visit is to present the loan details to the borrower, gets the borrower's acceptance to repay the loan. Training is then conducted for borrower (or borrowers in the case of group lending) in running the

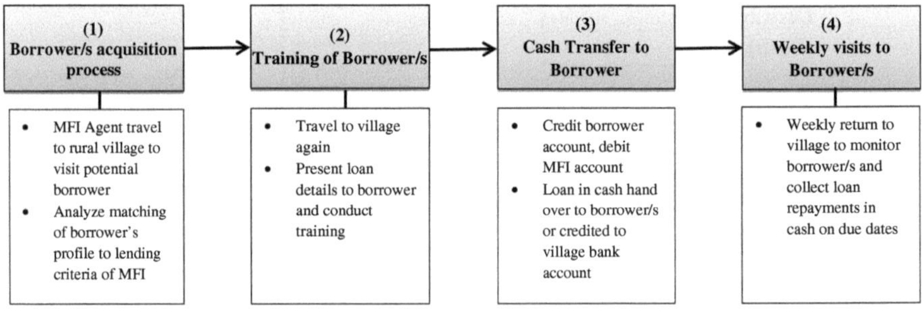

Fig. 2.3 The overall process of the pilot project

business, and matching the loan repayments to the business operations. The MFI agent writes a report of what has been done and the observations of the borrowers during the training. This process normally takes 5–7 days.

2.5.3 Cash Transfer to Borrower

The agent returns to the MFI Office to hand in the borrowers documents for safe keeping. There is also an accounting process to credit the borrower account and debit the MFI account. The MFI Agent will now return to the village with the cash. The loan in cash is handed over to the borrower/s, or in some rare cases the cash is credited to a village bank account of the borrower/s. This process takes 3–5 days.

2.5.4 Weekly Visits to Borrower/s

The MFI Agent returns to the village weekly to monitor the borrowers/s and collects the loan repayments in cash on due dates. The MFI agent writes a report on the ability of the borrower to run the business and also any observations on loan repayment ability. If cash, is collected the MFI agent shall return to the city with the cash and pass the cash to the MFI Accountant to pass accounting entries to the borrower's loan account and MFI's accounts. This process takes 3–5 days or even longer depending on the loan repayment ability of the borrower/s. The weekly visits will be done repeatedly until the loan is fully paid off.

There are few omissions to the loan process described above. Firstly, it has not taken into account of loan default or cash pilferage or adverse collusions to cheat the system. Secondly, in addition to the lending there could be credit insurance provided to borrowers. In such cases, where the loan defaults, the credit insurance company shall investigate and repay the loan on behalf of the borrower. That has not been articulated. Lastly, there could be savings tied to the loan. In such cases, it appears that there is some measure of "comfort" provided to the MFIs by the borrower.

2.6 Conclusion

In summary, for MFIs to scale fast enough to serve the huge demand of unbanked population and maintain financial sustainability, they require new business models and technology innovations to lower the interest burden by (1) lowering the operating cost, (2) accounting for legitimate activities and frauds, (3) lowering or mitigating risks of microcredits.

It was identified that the following processes can be replaced, fully or partially, by mobile phone applications thereby lowering the transaction and administrative costs:

1. Loan evaluation (with a credit scoring application).
2. Training (self-paced learning application).
3. Cash transfer for repayment and cash withdrawal or cash deposit in case of savings (with a peer-to-peer application).
4. Loan monitoring (transactions of borrowers can be monitored daily from transaction data on the backend server).
5. Deposit Savings can be topped-up to a mobile account.
6. Purchasing an credit insurance can be transacted through a mobile application.

Thus mobility technology offers much hope to emerging markets. It provides a new approach to lowering the transaction and administrative expenses of MFIs, that in turns lower cost of borrowing for a large majority of the low income population. In the end, it is envisaged that this will expand the social outreach and provide a broader range of financial services to the vast rural populations.

References

Ardic OP, Ibrahi JA, Mylenko N (2011) Consumer protection laws and regulations in deposit and loan services: a cross-country analysis with a new data set. Policy research working paper series

Askarany D, Yazdifar H (2007) Why ABC is not widely implemented? Int J Bus Res vii(1):93–98

BBC (2013) The Bangladesh poor selling organs to pay debts. http://www.bbc.com/news/world-asia-24128096

Bogdănoiu CL (2009) Activity based cost from the perspective of competitive advantage. J Appl Econ Sci (JAES) (7):5–11

CGAP (2008) Advancing financial access for the world's poor. CGAP 2008 Annual Report

Chaia A, Dalal A, Goland T, Gonzalez MJ, Morduch J, Schiff R (2009) Half the world is unbanked (Financial access initiative framing note). New York, Cambridge, New Haven

Chea AC (2011) Activity-based costing system in the service sector: a strategic approach for enhancing managerial decision making and competitiveness. Int J Bus Manage 6(11):3

Colgate M, Alexander N (2001) Retailers and diversification: the financial service dimension. J Retailing Consum Serv 9(1):1–11. doi:10.1016/S0969-6989(01)00009-1

de Aghion BA, Morduch J (2004) Microfinance—where do we stand? Financ Dev Econ Growth

Dickinson V, Lere JC (2003) Problems evaluating sales representative performance?: try activity-based costing. Ind Mark Manage 32(4):301–307. doi:10.1016/S0019-8501(02)00203-1

Dietrich A, Wanzenried G (2014) The determinants of commercial banking profitability in low-, middle-, and high-income countries. Q Rev Econ Financ 54(3):337–354. doi:10.1016/j.qref.2014.03.001

Estrin T, Kantor J, Albers D (1994) Is ABC suitable for your company?: here's an objective way to analyze your company's products and operations so you can determine the feasibility of implementing activity-based costing before wasting precious funds on a system that may be inappropriate. Manage Account-New York 75:40

Ghatak M (1999) Group lending, local information and peer selection. J Dev Econ 60(1):27–50. doi:10.1016/S0304-3878(99)00035-8

Goldmark L (2001) Microenterprise development in Latin America: towards a new flexibility. J Socio-Econ 30(2):145–149. doi:10.1016/S1053-5357(00)00095-0

Gosselin M (2006) A review of activity-based costing: technique, implementation, and consequences. In: Christopher S, Chapman AGH, Michael DS (eds) Handbooks of management accounting research, vol 2. Elsevier, New york pp 641–671. doi:10.1016/S1751-3243(06)02008-6

Hammond A, Prahalad CK (2002) Serving the world's poor, profitably. Harvard Bus Rev 80 (9):48–59

Haq M, Skully M, Pathan S (2010) Efficiency of microfinance institutions: a data envelopment analysis. Asia-Pac Financ Markets 17(1):63–97. doi:10.1007/s10690-009-9103-7

Heisinger K, Hoyle J (2014) Management accounting. Flat World Education, Inc

Hempel DJ (1976) Consumer satisfaction with the home buying process: conceptualization and measurement. Cambridge

Hermes N, Lensink R (2011) Microfinance: its impact, outreach, and sustainability. World Dev 39 (6):875–881. doi:10.1016/j.worlddev.2009.10.021

Hicks DT, Olejniczak EJI, Curell BA (2009) Measuring customer and product profitability at community and regional banks. J Perform Manage 22

Homburg C (2004) Improving activity-based costing heuristics by higher-level cost drivers. Eu J Oper Res 157(2):332–343. doi:10.1016/S0377-2217(03)00220-0

Horngren CT, Foster G (1987) Cost accounting: a managerial emphasis. Prentice-Hall, Englewood Cliffs

Hughes N, Lonie S (2007) M-PESA: mobile money for the "unbanked" turning cellphones into 24-hour tellers in Kenya. Innovations: Technol, Governance, Globalization 2(1–2):63–81. doi:10.1162/itgg.2007.2.1-2.63

Islam M (1995) Peer monitoring in the credit market. J Contemp Asia 26:452–465

James PC (2013) An analysis of the factors influencing the adoption of activity based costing (ABC) in the financial sector in Jamaica. Int J Bus Social Res 3(7):8

Kantor J, Maital S (1999) Measuring efficiency by product group: integrating dea with activity-based accounting in a large mideast bank. Interfaces 29(3):27–36. doi:10.1287/inte.29.3.27

Kaplan RS, Anderson SR (2004) Time-driven activity-based costing. Harvard Bus Rev 82 (11):131–140

Kee R, Schmidt C (2000) A comparative analysis of utilizing activity-based costing and the theory of constraints for making product-mix decisions. Int J Prod Econ 63(1):1–17. doi:10.1016/S0925-5273(99)00005-5

Khataie AH, Bulgak AA, Segovia JJ (2011) Activity-based costing and management applied in a hybrid decision support system for order management. Decis Support Syst 52(1):142–156. doi:10.1016/j.dss.2011.06.003

Kotler P (1976) Marketing management: analysis, planning and control. Prentice-Hall International, Englewood Cliffs

Koveos P, Randhawa D (2004) Financial services for the poor: assessing microfinance institutions. Manag Financ 30(9):70–95. doi:10.1108/03074350410769281

KWFT (2014) Kenya Women Microfinance Bank. http://www.kwftdtm.com/

Lafourcade A-L, Isern J, Mwangi P, Brown M (2006) Overview of the outreach and financial performance of microfinance institutions in Africa. MicroBanking Bull

Lee JY (2002) An examination of international differences in adoption and theory development of activity-based costing. Adv Int Acc 15:65–77. doi:10.1016/S0897-3660(02)15019-7

Liu LYJ, Pan F (2007) The implementation of activity-based costing in China: an innovation action research approach. British Acc Rev 39(3):249–264. doi:10.1016/j.bar.2007.05.003

Nenovski T, Jolevska ED, Msc IA (2012) Banking services in terms of changing environment: the case of Macedonia. Procedia—Social Behav Sci 44:347–356. doi:10.1016/j.sbspro.2012.05.038

Qian L, Ben-Arieh D (2008) Parametric cost estimation based on activity-based costing: a case study for design and development of rotational parts. Int J Prod Econ 113(2):805–818. doi:10.1016/j.ijpe.2007.08.010

Rasiah D (2011) Why activity based costing (ABC) is still tagging behind the traditional costing in Malaysia? J Appl Financ Banking 1(1):83–106
Robinson M (2001) The microfinance revolution: sustainable finance for the poor. The World Bank, Washington, DC
Rosenberg R, Gonzalez A, Narain S (2009) Are microcredit interest rates excessive. World Bank, Washington, DC
Rosenberg R, Gaul S, Ford W, Tomilova O (2013) Microcredit interest rates and their determinants (2004-2011). CGAP
Roztocki N (2005) A procedure for smooth implementation of activity based costing in small companies. Eng Manage J 16:19
Schoute M (2011) The relationship between product diversity, usage of advanced manufacturing technologies and activity-based costing adoption. British Account Rev 43(2):120–134. doi:10.1016/j.bar.2011.02.002
Sena M (2010) Make or buy: a comparative assessment of organizations that develop software internally versus those that purchase software. In: Conference on information systems applied research proceedings
Shankar S (2007) Transaction costs in group microcredit in India. Manag Decis 45(8):1331–1342
van Raaij EM, Vernooij MJA, van Triest S (2003) The implementation of customer profitability analysis: a case study. Ind Mark Manage 32(7):573–583. doi:10.1016/S0019-8501(03)00006-3
Van Tassel E (1999) Group lending under asymmetric information. J Dev Econ 60(1):3–25. doi:10.1016/S0304-3878(99)00034-6
Vong J (1994) Strategic response quotient—a major indicator of bank performance. SES J
Vong J, Fang J, Insu S (2012) Delivering financial services through mobile phone technology: a pilot study on impact of mobile money service on micro–entrepreneurs in rural Cambodia. Int J Inform Syst Change Manag 6(2):177–186
Wegmann G (2011) Activity-based management in france: a focus on the information systems department of a bank. In: Proceedings of the EBMM 2011 international conference
Wills G, Day A (1984) Marketing and selling at work: the IMCB/NatWest management development programme. Int J Bank Mark 2(1):3–11. doi:10.1108/eb010730
Woolcock M (2001) Microenterprise and social capital: a framework for theory, research and policy. J Behav Exp Econ 30(2):193–198
Yunus M, Jolis A (1999) Banker to the poor: micro-lending and the battle against world poverty. Public Affairs, New York

Chapter 3
Bank Ratings in Emerging Asia—Methodology, Information and Technology

Abstract Bank rating is as old as bank regulatory authorities. The most common approach to bank rating is on-site examination. The bank supervision authority conducts an audit of physical bank records and documents and other observations. In particular, the on-site examination looks at the bank's assets, capital structure, management ability and professionalism of the bank practices. A summary rating is assigned based on this examination. This approach was reported in 1974 in a publication of the Federal Reserve Bank of New York. This rating system was called into question for the first time during the Asian Financial Crisis (AFC) of the 1997 that affected South Korea, Thailand, and Indonesia. IMF painfully restructured the South Korean economy. Thailand suffered a severe outflow of funds and banks collapsed. The entire Indonesian banking system had to be revamped. Then the global financial crisis struck in 2007–2008, where the western world was hit harder than the AFC that led to the collapse of venerable financial institutions such as Lehman Bros and Bear Sterns. This begs the question: why bank ratings did not, or cannot, forewarn imminent failures in the banking system. In 2009, it was reported that banks were usually rated higher if they had greater capitalization, larger assets, and higher return on assets and equity. In 2013, in a most scathing attack on bank ratings, it stated that most failing banks had enjoyed investment status ratings just before they defaulted. This paper revisits the age-old CAMEL model and its derivatives, and a more recent EAGLES framework that alerted the Asian Financial Crisis of the 1990s. Based on the experiences of two crises in the last two decades, what lessons did we learned, especially on the early warning signals, contagion effect, information transparency and the availability of technology processes to collect and harness critical information.

Keywords Bank rating · CAMEL · EAGLE

J. Vong and I. Song, *Emerging Technologies for Emerging Markets*,
Topics in Intelligent Engineering and Informatics 11,
DOI 10.1007/978-981-287-347-7_3

3.1 Introduction

Traditionally the bank regulatory authorities assess the five key aspects of the operations of a bank—capital, assets, management, earnings and liquidity (CAMEL). An overall CAMEL rating is derived from giving a score from one (for the best) to five (worst rating) is given to each of the five factors. The five composite rating levels are described by the Commercial Bank Examination Manual written by the Board of Governor of the Federal Reserve System in the United States (Dzeawuni and Tanko 2008; Roman and Şargu 2013).

1. Best rating; sound in every respect
2. Fundamentally sound but has modest weaknesses
3. Financial, operational, or compliance weaknesses that give cause for supervisory concern
4. Serious financial weaknesses that could impair future viability
5. Worst rating; Critical financial weaknesses that indicate the probability of failure extremely high in the near term. Requires constant supervision.

3.2 CAMEL—The Vintage Model

3.2.1 Capital

The capital adequacy requirement has profound impact upon the banking industry. The international guideline that stipulates a bank must have a minimum capital equivalent to 8 % of risk adjusted assets. The higher the level of capital adequacy ratio (CAR) means that the bank has stronger capacity to withstand unexpected losses.

The principal objective of a bank's capital is to ensure that an adequate level of capital is maintained to bear any losses that may result from credit defaults or losses. The risk-based capital adequacy framework is consistent with international standards of the Basel Committee. The Basel regulation has three aspects: capital base element (commonly known as Tier 1 and Tier 2), the required capitalization level, and the risk weighted regime.

As stated previously, the capital base has two elements namely Tier 1 and Tier 2 capital. The T1 ratio is equal to the core capital divided by risk-adjusted assets. And this should be at least 4 %. Both T1 and T2 capital must be at least 8 % of risk weighted assets. The risk weights assigned to assets ranges from 10 to 100 % (Bank for International Settlements 2012).

3.2.2 Asset Quality

The usual approach to assess loan quality is to have an on-site inspection of the bank's loan books. As this is not always possible, asset quality can be measured by the level of provisions made, that is, bad and doubtful debts (BDD) provision as a percentage of total loans. Taking a conservative stance will dictate that the quantum of provision to be estimated as higher rather than lower. This provision is usually measured as a percentage of loans and other credit receivables. Therefore, the banks must manage their loans levels closely to given capital, constraints, and risks preferences.

3.2.3 Management

This is where financial supervision auditors would examine the soundness of the policies, regulations and decisions taken by the bank and rate the bank on a scale of 1–5. The most disconcerting issue appears to be the assessment of bank management quality. This subjective factor has defied measurement in the last 40 years. Researchers have alluded to soundness of management practice by examining the decisions made by the senior management of the bank. It is often thought that good management can result in stable returns, significant market, increasing asset growth, or increased in equity value. If these are reliable indicators, then why did creditable banks with well-established policies and experienced managers collapsed in the 2007–2008 global financial crisis? Were there any signs that these banks did not adhere to good management practices? How did the banks managed to escape the radar of financial supervision authorities and international rating agencies?

3.2.4 Earning

In measuring earning ability there are three noteworthy indicators—return on Assets (ROA), return on equity (ROE) and Income-Cost ratio, also known as income to overheads ratio. The ROA is important to bank management to monitor their efficiency in asset acquisition, while the ROE indicates the level of return to investors holding an equity stake in the bank. The income to overheads ratio provides the extent to which income covers overheads. Higher coverage indicates that the bank has better earning capacity. Overheads tend to grow with the level of staffing, while income trends follow the market economy.

3.2.5 Liquidity

Liquidity can be described as the ability of a bank to have sufficient funds to meet cash demands for loans, deposit withdrawals and operating expenses. For this reason, a balance should be found between the amount of deposits garnered and the quantum of loans extended. The indicator is the deposit-to-loans ratio.

3.2.6 CAMEL Scoring

Table 3.1 describes how banks can be rated 1–5 based on their financial performance under the CAMEL framework.

However, the CAMEL approach to bank analysis is contingent to areas of indeterminacy, subjectivity and sometimes inconsistency (Cole and Gunther 2008). As most bank analysts and examiners would acknowledge, there are instances when an examination of the bank physical documents and accounting records cannot decide whether to give an average or below average score. The 'good' score of 1, and the 'bad' score of 5, may be easier to spot, but the 'in-betweens' like 2, 3 and 4, may be more difficult. At the conclusion of the bank examination, a 'best practice' bank may receive a score of 1 for each of the CAMEL indicators. By assigning the scores, the bank examiner falls into the trap of subjectivity. Take for example, the ratios and ratings tabulated as Table 3.1 CAMEL Indicators (Ferrouhi 2014; Nurazi and Evans 2005; Rozzani and Rahman 2013).

Minor improvements have since been made, such as CAMELOT, which included operating environment and transparency, and the CAMELS, where strategy issues are considered too (Alber 2014; Munir et al. 2011). But it still leaves much to be desired as the subjectivity element remains.

Table 3.1 CAMEL indicators

Category	Ratio	Score 1 (%)	Score 2 (%)	Score 3 (%)	Score 4 (%)	Score 5 (%)
Capital adequacy ratio	Equity capital/total assets	Above 11	8–11	4–8	1–4	Below 1
	Equity capital—fixed asset/ total loan + securities					
Assets quality	Earnings before in-come tax/productive assets	Below 1.5	1.5–3.5	3.5–7	7–9.5	Above 9.5
	Off-balance sheet activities/equity capital					
Management	Net income/total assets	Below 25	30–26	38–31	45–39	Above 46
	Net income—interest expense/total assets					
Earnings	Earnings after tax/ operating income or sales	Above 1.50	1.25–1.50	1.01–1.25	0.75–1.00	Below 0.75
Liquidity	Total loan/total deposit	Below 60	60–65	65–70	70–80	Below 80

Table 3.2 CAMEL ratings of selected indonesian banks (October 1996) published 6 months before the financial turmoil

	Bank Bali	Bank Danamon	Bank International	Bank Niaga
Overall rating	Investment grade	Investment grade	Investment grade	Investment grade
Capital (CAR %)	11.8 %	10.3	9.3	9.5
Asset quality (NPL %)	1.9	1.9	1.8	2.7
Earning (ROE %)	20.7	13.0	23.4	20.7
Management	6th largest private bank	2nd largest private bank	4th largest private bank	7th largest private bank
Liquidity (LDR %)	82.0	95.7	73.9	91.1

Source Li et al. (1998), Vong and Tham (1998)

Table 3.3 EAGLES indicators

EAGLES	Financial indicators
Earning	Return on asset (ROA); return on equity (ROE); interest margin %; income to cost ratio %
Asset quality	Non-performing loans %; loans classification
Growth	Loan growth %; deposit growth %; NPL growth %
Liquidity	Loans to deposit %
Equity	Capital adequacy ratio %
Strategic response	Interest burden; number of times cover

The CAMEL evaluation framework used by an international investment bank to assess five four banks in Indonesia approximately 12 months before the financial collapse is shown in Tables 3.2 and 3.3.

3.3 EAGLES—The Emerging Model

The EAGLES benchmark monitors bank performance from the perspectives of earning ability, asset quality, growth, liquidity, equity and strategy. The focus is not entirely dissimilar to the CAMEL. However there are three distinct differences that enable a clearer assessment of the condition of the bank. Firstly, EAGLES is conducting a financial analysis of the bank. In other words, if the Return on Asset is 1.75 %, it is noted as 1.75 % and not given a score of 1–5 (as in the CAMEL framework). The EAGLES assessment consists of financial ratios, and not subjective ratings of 1–5. Needless to say, if a trend analysis can be performed on these financial ratios over consecutive time periods, the imminent weakness or failures of banks or the banking system can be easily discerned.

Secondly, the bank management quality is determined by the pricing of the deposit rates and loan rates, managing non-interest operating costs, and garnering fee income. These four activities are important because it determines bank management's ability in deposit taking, lending, obtaining fee-based income, and controlling overheads (Vong 1994). The strategic response quotient (SRQ) considers four key financial information of the bank, that is, interest income (ITI), interest cost (ITC), non-interest income (NII) and non-interest cost (OHC). It is measured by taking the interest margin and dividing that by the net operating cost (i.e. fee income less non-interest expense), as shown below:

$$\frac{(ITI - ITC)}{(NII - OHC)}$$

As to what is an appropriate SRQ level will depend on the bank's strategic positioning. A high SRQ could imply ample fee income, or low staff cost, or high net interest income. It measures the number of times the net interest income covers the interest burden. From the SRQ, a bank examiner is able to discern the extent of the credit and operational risks to which bank is subjected.

Thirdly, the growth rates of lending and deposit taking, and interdependent financial indicators are compared. For example, is the non-performing loans growing as fast as the loan growth, and why or why not? Is deposits growth in line with the loan growth?

In comparing Tables 3.1 (CAMEL Ratings) and 3.3 (EAGLES Ratings), the following differences can be found: (i) the EAGLES model has the "Growth" indicator that tracks growth of loans, deposits, and non-performing loans; (ii) the EAGLES model replaces 'Management,' under the CAMEL framework with the Strategic Response Quotient (SRQ) which is an objective measure derived from the number of times the net interest income covers the interest burden.

Considering EAGLES financial analysis of four Indonesian banks in Table 3.4, there are two important observations to consider. Firstly, it was observed that the asset is growing significantly faster than the provisioning for non-performing loans. This under-provision is a sign that the bank taking too much credit risk and under

Table 3.4 EAGLES financial analysis (as at end 1996)

EAGLES	Bank Bali	Bank Danamon	Bank International	Bank Niaga
Earning (ROE %)	20.7	13.0	23.4	20.7
Asset quality (NPL %)	1.8*	1.1*	2.4*	1.5*
Growth (Loans %)	26.2*	40.9*	21.8*	26.9*
Liquidity (LDR %)	82.0	95.7	73.9	91.1
Equity (CAR %)	11.8 %	10.3	9.3	9.5
Strategic response (times cover)	2.5	2.0	4.1	1.9

*The asset growth was more rapidly than the provisioning of non-performing loans

Source Li et al. (1998), Vong and Tham (1998)

estimating the inherent dangers. Secondly, although the interest burden was adequately covered, the bank could be charging a much higher loan interest rate than the market rate. It is commonly acknowledged that customers who pay higher loan rates are usually those with higher credit risks.

It was precisely these 2 factors that alerted the authors of an impending collapse of the banks in Indonesia and Thailand (Vong and Tham 1998).

3.4 Lessons—Two Crises, Three Continents (1997–2007)

Two financial crises, over a decade apart, over two or three different continents have certainly shed some light on the frailties of the banking system and vagaries of the financial markets. It is good to illuminate some of these lessons.

3.4.1 Lesson 1: Strong Macroeconomic Fundamentals Are Critical to Maintain the Confidence of the Investors

Excess global liquidity, inadequate buffers, and indiscriminate lending, are not signs of strong macroeconomic policies (Wellink 2011; Vong and Tham 1998). The financial supervision authorities and the market regulators should be watching the market movements and bank indicators like a hawk. This is even true when the mood of market seems excellent. It has proven on the two banking failures that the market conditions look healthy enough 6–12 months before disaster struck.

In US, the crisis was related to the securitization residential mortgages that were given based on weak loan repayment criteria (often referred to subprime mortgages). The failure of many borrowers to repay has adversely affected the value of these mortgages. In turn, this led to a fall in securities price that meant spreading risk to other parts of the financial system. A contagion effect took over the markets. Eventually with the downward spiral that resulted in the global financial markets.

In Asia, the crisis was first felt in Thailand and then spread to Indonesia and other parts of Asia. It all began with a loose monetary policy and a credit boom. Foreign banks were lending cheaper US dollars to domestic banks. In turn the domestic banks advised the domestic borrowers to take US dollar loans because the US dollar had lower interest rates, compared to local currencies. The foreign currency loans were exposed to exchange rate risk, because many loans were not hedged. Thus when the foreign banks suddenly withdrew the US dollar loans from the market, the domestic banks could not enforce payments from their domestic borrowers. At the same time, the value of the domestic currencies lost ground against the US dollar.

This left the borrowers with the one choice, to default rather than repay.

3.4.2 Lesson 2: There Must Be Strong Domestic Safeguards to Help Neighboring Countries to Prevent the Contagion Effect

Research has shown that the greater the financial contagion, the greater the severity of the financial crisis for the domestic economy (Ozkan and Unsal 2012). Regression analysis shows that a country is more vulnerable to contagion if it has a more leveraged banking system, greater trade exposure, weaker macroeconomic fundamentals, and larger international portfolio investment liabilities. Countries are less vulnerable, however, if they have larger international financial portfolio investment assets (which can provide a buffer against shocks) and are less reliant on debt (versus equity) for international financing (Forbes 2012). Therefore, to minimize contagion risks, structural reforms have to be in place before a negative shock occurs. It is also recommended that priority should be given to reducing the debt-equity ratio in banking system. To balance international portfolio liabilities, countries should support efforts to diversify and invest abroad.

3.4.3 Lessons 3: Governments, Financial Institutions, and Corporations Must Communicate Well with Investors, Creditors, and to the Market

In both the financial crises there seems to be decision flips-flops, unclear messages, refusal to admit failure, political in-fighting, and lack a clear direction to stabilize the market. In the Asian financial crisis, both Thailand and Indonesia were forewarned months ahead of the market collapse (Vong and Tham 1998). In the US market failure, Lehman Bros was put on watch list months before the government stepped in. There seems to be a period of denial, of which that time could have been applied to take precautionary measure. Not surprisingly, even Bear Sterns, now defunct, was recognized as the "Most Admired" securities firm in Fortune magazine's "America's Most Admired Companies" survey of 2005 (Forbes 2012). The annual survey is a prestigious ranking of employee talent, quality of management and business innovation. Nobody knew it would be bankrupt by March 2008.

This begs the question: given the vast amount of data collected and experiential learning since the 1930 Great Depression, stock market collapses and bank failures, surely the approaches to evaluate banks have improved?

Table 3.5 Integrated MIS of a financial regulator

An Integrated MIS	
Internal information subsystem	External information subsystem
Demand deposits	Capital flows
Savings deposits	Financial markets
Time deposits	Market trends
Money market	Foreign exchange
Imports/exports	Competitive position
Loans	Social trends
Financial performance	Actual versus forecast versus history
Human resource data	
Actual versus forecast versus history	

Source Vong (1986)

3.5 Technology to Provide Pertinent Information

System and technology must be designed and implemented to track performance of banks against the backdrop of the economic situation and other macroeconomic indicators. There should be a mechanism for economic surveillance to detect areas of risks and provide early warning signals so that financial disaster can be mitigated. This mechanism can be integrated with the monitoring systems of the financial supervision and regulation authorities. This system integration, in turn, will enhance the collaborative wisdom and strengthen the capacity of the financial regulation authorities to respond to financial calamities.

With the vast amount data, it is possible to perform data analytics to detect market sentiments; identify, track and analyze early warning signals; test for credit deterioration; measure risk; perform scenario analysis; conduct stress testing (Analytics 2010).

The investment community and bank regulators need a high level of information disclosure to create a healthy and buoyant financial market (Vong and Tham 1998). While the investment community does not have sufficient information, most regulators have the information. They should manage the data better to formulate effective surveillance.

Table 3.5 shows a possible Integrated Management Information System (MIS) for a financial regulator to systematically conduct economic and financial surveillance. With such an integrated system it is possible to evaluate bank performance using CAMEL and EAGLES indicators alongside updated information on financial markets and the economy on a dashboard.

References

Alber N (2014) The effects of banking regulation on asset quality: a panel analysis. Int Bus Res 7 (7):164–174
Analytics Ms (2010) Credit research and risk measurement
Bank for International Settlements International regulatory framework for banks (Basel III) (2012) http://www.bis.org/bcbs/basel3.htm
Cole RA, Gunther JW (2008) A CAMEL rating's shelf life. MPRA Paper
Dzeawuni WA Tanko MI (2008) CAMELs and Banks Performance Evaluation, The WayForward
Ferrouhi EM (2014) Moroccan banks analysis using CAMEL mode. Int J Econ Fin Issues 4 (3):622–647
Forbes KJ (2012) The big "C": identifying and mitigating contagion. Paper prepared for 2012 Jackson hole symposium hosted by the Federal Reserve Bank of Kansas
Li WJ, Liu L, Tong WL (1998) Indonesian Banking; Marriage or Perish. In: Tan TM, Low AM, Williams JJ, Choo TM (eds) Business opportunities in Indonesia. Prentice-Hall, Singapore, pp 105–128
Munir R, Perera S, Baird K (2011) An analytical framework to examine changes in performance measurement systems within the banking sector. Australas Acc Bus Financ J 5(1):93–115
Nurazi R, Evans M (2005) An Indonesian study of the use of CAMEL(S) ratios as predictors of bank failure. J Econ Soc Policy 10(1):143–167
Ozkan FG, Unsal DF (2012) Global financial crisis, financial contagion and emerging markets. IMF working paper
Roman A, Şargu AC (2013) Analysing the financial soundness of the commercial banks in Romania: an approach based on the camels framework. Procedia economics and finance 6 (0):703–712. doi:10.1016/S2212-5671(13)00192-5
Rozzani N, Rahman RA (2013) Camels and performance evaluation of banks in Malaysia: conventional versus Islamic. J Islamic Financ Bus Res 2(1):36–45
Vong J (1986) Information systems for planning and control. Manag Decis 24(5):17–19
Vong J (1994) Strategic response quotient—a major indicator of bank performance. SES J
Vong J, Tham KC (1998) Thailand and Indonesia beat the brunt: what went wrong? Banker's J Malays 104:4–5
Wellink N (2011) Basel III and beyond. FSI and EMEAP working group on banking supervision

Chapter 4
Microfinance and Gender Equality in Indonesia

Abstract The financial inclusion of women micro entrepreneurs is both important and necessary for the holistic socio-economic development of a nation. This paper aims to address the gap in the study of the processes of financial inclusion by women micro-entrepreneurs in Indonesia. The research will contribute to the pool of academic resources related to financial inclusion, micro credit, and women entrepreneurship that intend to improve the processes and benchmarking for financial inclusion initiatives. In writing this paper, there was extensive literature research plus in situ research conducted in Indonesia. There is also a significant amount of evidence that is related to factors that contribute to gender inequality. Both desk research and field research found that financial exclusion is linked to financial education, cultural norms of rural Indonesia and high banking cost and mobile payments. In conclusion, the authors recommend improving processes in designing financial products and financial education, and lowering the cost of banking, especially in the use of mobile payment technology, and instituting micro-insurance. Further research is recommended to generate hypotheses relating to financial inclusion and women entrepreneurship in Indonesia.

Keywords Financial inclusion · Financial exclusion · Cost of banking · Bank products · Education for women · Micro finance · Rural indonesia · Women micro-entrepreneurs · Payment technology · Gender inequality · Access to formal finance

4.1 Introduction

Over the last 10 years, Indonesia has been the largest economy in Southeast Asia with a GDP growth of between 4.5 and 6.2 %. Although the nation is facing multiple problems, such as corruption and instability, Indonesia has emerged from the global financial crisis with a surprisingly positive reputation. In 2010, it was called a "golden child" (Vong 2010). This is good news for small and medium-sized enterprises (SMEs) and it is expected that there will be a significant pick up in SME Banking.

J. Vong and I. Song, *Emerging Technologies for Emerging Markets*,
Topics in Intelligent Engineering and Informatics 11,
DOI 10.1007/978-981-287-347-7_4

The microenterprises and SMEs play an important role in Indonesia. In the 2006 Census done by Indonesian government, SMEs constituted 99 % of all businesses in the country, employed 97 % of the work force, and accounted for 54 % of that year's GDP.

Table 4.1 shows an Economic Census figure between 1996 and 2006. It shows that the SME sector grew 38 % in terms of firms and 46 % in terms of employment. In 2006, there were nearly 23 million firms in the country with nearly 50 million workers.

The small business-lending (SBL) segment probably holds the greatest promise for higher returns and operational economies of scale for banks in Indonesia. The Indonesia 2006 Census estimates that there are about 3.6 million small enterprises. According to (International Finance Corporation 2010a), 90 % of the banks agreed that this segment has higher profitability than commercial banking, and is less competitive than retail banking as shown in Table 4.2. It is also surveyed that 20 % of small enterprises (or 727,000) intend to borrow this year.

Table 4.1 Distribution of firms among the micro, small, medium and large categories (based on firms' turnover)

SME category	Distribution (%)	Estimated numbers
Micro	83.7	About 19 million firms
Small	15.8	About 3.6 million firms
Medium	0.7	About 161,000 firms
Large	0.2	About 46,000 firms

Table 4.2 Key statistics of SMEs and women in Indonesia

Micro and SMEs account for 54 % GDP 2006 (i.e. IDR 3338 trillion), employs 97 % of the workforce, and accounts for 54 % of GDP
23 million SMEs constitute 16.4 % of all firms in Indonesia
Number of firms grew 38 % (1996–2006)
0.7 % (160,000) medium-sized enterprises: 15.8 % (3.6 million) and 83.7 % (19 million) micro
20 % of small business (727,000 firms) intend to borrow in 2010
15 % expressed interest in internet-based fin products and property insurance products
70 % are sole proprietorships
About 49 % of SMEs wish to borrow up to IDR 150 million
86 % of SMEs have up to IDR 50 million revenues per month
28 % of SMEs are women-owned of which 97 % are outside Jakarta
90 % Women owned SME use their own savings for funding the business
"Best lenders" are BRI and BCA
Margin erosion at corporate banking has moved banks to commercial lending to "medium sized firms"

Source International Finance Corporation (2010b)

The development of microenterprises are considered to be instrumental in providing employment and raising the standard of living of poor people all over the world (Singh et al. 1999), however, it concurs with research literature that reports a higher percentage of micro-entrepreneurial activity is reflective of higher poverty 'distress', in reference to Indonesia (Tambunan 2011). The nation is noted to take firm steps to counter poverty levels in the rural areas. This may be evident from the number of micro-enterprises undertaken by poor households as they are "pushed" to take on activities to incur their primary and secondary income. During the Asian financial crisis of 1997, microenterprise activity in Indonesia absorbed laid-off workers and presently it represents 90 % of all Indonesian firms (Vial 2011).

In 2010, according to the IFC survey report, financial access is highly skewed to urban areas whereas only 20–34 % of rural households had access to banking services (International Finance Corporation 2010a). Another report states that out of 230 million Indonesians only 55 million have bank accounts (Vong 2010). Based on the analysis highlighted above we can assume that majority of the unbanked population would be rural women due to the several constraints. This means there is significant scope for financial inclusion of rural women entrepreneurs. The challenge that micro-entrepreneurs currently face is that there are not enough banks located in close proximity to remote rural areas. The opportunity cost is very high for rural women to travel to the banks in cities (Columbia Business School 2012). Hence the first step for financial inclusion process is to provide easy access with minimum opportunity cost. Most importantly, banks need to provide a full range of financial services, as customers would like to deal with only one bank account for all their requirements (Vong 2010). This can be followed by minimized documentation process, female oriented services, and products.

4.2 Microcredit and Micro-entrepreneurs in Indonesia

The microfinance institutions in the earlier decade gave more emphasis on microenterprise lending, above all other services (Eversole 2003). However, the strategy has changed towards offering a range of non-financial business development services such as enterprise management, accounting courses, marketing assistance etc. This is done to support more sustainable development of the economic structure.

It can be deducted that there is insufficient competition in the micro financing sector that has led them to ignore the needs of various niche target markets (Tambunan 2009). This may have also hampered the creation of innovative financial products for these particular targets in question. According to Badan Pusak Statistik (aka Statistical Office of Indonesia) resources, the micro-enterprises located in rural/backward areas remarked of not receiving any credit from banks or other financial institutions. The absence of formal assistance required them to raise capital from their own savings, family members, and informal money lenders (Tambunan 2011; Vong 2010).

International Finance Corporation (IFC) has been playing a catalyst role with its program called "Banking on Women", with financial institutions and banking partners all over the world. The main focus of IFC's program is to ensure that businesses run by women entrepreneurs are profitable and sustainable. This plan is in line with the Millennium Development Goals (MDG) and economic growth of developing countries. Approximately, 30 % of the businesses registered across the world are women-owned entities. However, only 5–10 % of these entities have access to loans from commercial banks and financial institutions. Some of the impediments to women entrepreneurs from accessing finance, as listed by IFC are —weak management and business skills, unfavourable business and regulatory environment, lack of tailor-made banking products and financial services for women (Vong 2010).

The products and investment services of IFC are:

1. Risk sharing facilities with banking and non-banking partners
2. Equity/capital
3. Credit insurance
4. Loans/credit lines
5. Supply chain finance and distributor finance

IFC also offers Advisory Services so that financial institutions can reach out more effectively to women entrepreneurs. These services include: staff training, segmentation or market positioning, strategic planning and product repositioning. It trains women entrepreneurs in business management and planning, financial literacy, and elucidate credit/bank application process. In 2010, as a part of Banking on Women program IFC had invested in Bank International Indonesia and has planned for many such investments in the coming years (International Finance Corporation 2013).

According to a survey conducted by Hearts and Wallets (Kuehn 2012), 49 % of women admitted that they are very inexperienced in financial services and their lack of understanding of financial products causes difficulty for them to formulate sound financial decisions. This has been confirmed through another study done by Prudential.

Research papers by Johar et al. (2006) and Vial (2011) suggested that there are informational asymmetries in Indonesia. They also mentioned that, before any discussion about the improvement of credit facility or financial services, it is important to test the financial knowledge and willingness of people to borrow. The tests and studies conducted in the researches were based on regression analysis using different factor combinations to look out for changes in results i.e. propensity to borrow, with each combination. The analyses revealed interesting results—the tendency to borrow is higher for women from female-headed households. Moreover, a strong correlation was found between education level of women and the size of loan received. They discovered that social network albeit increases the chances of women getting a loan, the education level decide the size of loan granted. They also identified that women's borrowing activity is not affected much by money lenders' policies.

A study in the United Kingdom (UK) has also strongly supported the hypothesis that, similar to Indonesia, women are more likely to perceive financial barriers to business start-up than men, ceteris paribus. Data suggests that in the UK, women are 7.4 % more likely than men to perceive financial barriers. However, there is empirical evidence given in the report that the level of education also plays a vital role in financial access to women. However, it does not provide enough supporting reasons for the negative perceptions of UK women related to financial barriers (Roper and Scott 2009).

4.3 Women Micro-entrepreneurs

According to Tambunan (2011), there was an up-rise in the positive international speculation to the concept of women's entrepreneurial participation. This has, especially in rural areas, invoked interest among policy makers, academics, and practitioners in Indonesia. It is looked upon as an opportunity to encourage mobilization of women talent as entrepreneurs to improve the rural economic conditions. Tambunan (2009) finds contrast in literature reviews about the predominant factors for entrepreneurship among women in rural Indonesia to that of more developed economies such as the United States and Korea. The former is found to, most often, be in an economic necessity to improve their family's economic status, whereas the latter desires to achieve fulfilment, independence, self-esteem, and liberty of choices. It is essentially true in countries with higher levels of poverty that the women seemed more to be 'pushed' towards entrepreneurship, rather than 'pulled' (Davis and Abdiyeva 2012). Davis and Abdiyeva (2012) found women to be confined to managing very small enterprises; mostly microenterprise with less than 10 employees. This is further supported with the similar cases in many countries such as Jordan, Botswana, Kenya, Lesotho, Malawi, Swaziland, and Malaysia.

In Indonesia, Tambunan (2011) attributes four main factors which have led to the low representation of women entrepreneurs: (1) Low educational and training opportunities, (2) heavy household chores, (3) legal, traditions, customs, and cultural norms in social context for Indonesian women, (4) lack of access to formal credit and financial institutions. He also mentions that these same factors have been observed to have changes for women in urban areas in the last 30 years. Through an interview in person, the author also deduced that majority of women entrepreneurs in manufacturing industry were mostly found to originate from the western and more developed regions. This gives further reason to investigate the opportunities offered differing to females hailing from urban societies in comparison to rural societies of Indonesia.

The various constraints detailed in the literature review points to the lack of access to financial credit. Kantor (2005), conducted an empirical analysis which confirmed that access of credit did not have statistically significant effect on the economic performance of the enterprise, but the size of the effect is still considered large and positively supportive for the importance of credit for micro entrepreneurs.

4.4 Constraints Suffered by Indonesian Women Entrepreneurs

According to the 2009 United Nations Centre for Human Settlements report (Narain 2009), it was found that 75 % of women in the world were discriminated in the formal and informal credit markets. For instance, Davis and Abdiyeva (2012) inferred from a survey conducted in Kazakhstan that 36 % of the women entrepreneurs interviewed still find difficulty to obtain formal bank finances for their enterprises over the past decade. This is due to their lack of permanent employment, lack of asset (land or housing) ownership to offer as security or the country laws acknowledgment to identify women as a minority. In general terms, it has been mentioned that there are various constraints faced by small entrepreneurs in Indonesia. This may alter according to region, sector or by the nature of enterprise. The constraints may enlist the lack of capital, access to supply materials, access to relevant business statistics, lack of marketing knowledge and product distribution (Tambunan 2011).

Tambunan (2011) mentioned that 'micro entrepreneurial finance' may be viewed as a risky investment for formal institutions because it often entail requests for non-collateralised forms of lending. Even on acceptance of the micro-loan, it may be inadequate and can particularly constrain the necessary supply of capital for micro entrepreneurs. He also extracted statistical data from Badan Pusak Statistik resources shown in Table 4.3. Through a pilot study, the author succeeded to uncover the main constraints in manufacturing industry. With focus on Micro-Enterprises (MIEs), 75 % have admitted to the presence of constraints with respect to capital, raw materials, and marketing.

Table 4.3 Main problems faced by SEs and MIEs in manufacturing industry (Tambunan 2011)

	SEs	MIEs	Total SEs and MIEs
Have no problem	46,485 (19.48)	627,650 (25.21)	674.135 (24.71)
Have problem	192,097 (80.52)	1,862,468 (74.79)	2,054,565 (75.29)
Raw material	20,362 (10.60)	400,915 (21.53)	421,277 (20.50)
Marketing	77,175 (40.18)	552,231 (29.65)	629,406 (30.63)
Capital	71,001 (39.96)	643, 628 (34.56)	714, 629 (34.78)
Transportation/distribution	5,027 (2.62)	49,918 (2.68)	54,945 (2.67)
Energy	40,605 (2.4)	50,815 (2.73)	55,420 (2.7)
Labor cost	2,335 (1.22)	14,315 (0.77)	16,650 (0.81)
Others	11,592 (6.04)	150,646 (8.09)	162,238 (7.90)
Total	238,582 (100.00)	2,490,118 (100.00)	2,728,700 (100.00)

4.5 The Result of Financial Exclusion of Women Entrepreneurs in Indonesia

According to the data produced from World Economic Forum (WEF), Indonesia has shown poor results in Gender Gap Index (GGI) with a rating of 0.6591 in 2012 (Hausmann et al. 2012). Its ranking dipped from 68th position in 2006 to 97th in 2012 out of 135 countries. As seen in Tables 4.4 and 4.5, this index is based on four main criteria of inequality between men and women: (1) Economic Contribution; (2) Education Opportunity; (3) Health and survival; and (4) Political Empowerment. The scale consists of index values, where zero represents inequality and one represents equality. This index measures the gender-based gaps in access to opportunities and resources in the respective countries but not the actual levels of resources and opportunities available in those countries.

Another index supporting the evidence of Gender gap is Gender Equity Index (GEI) developed by Social Watch, which states the value hundred indicates the least gender gap and zero indicated the highest gender gap. It classifies countries relevant to gender inequity based on 3 areas: Education, Economy activity and Women Empowerment. It states that Indonesia scored only 55 in 2010. However, the scores have improved to 62 in 2012. But it is still far from developed neighbouring countries like Singapore. Table 4.6 shows the GEI for ASEAN countries in 2012.

According to the Human Development Report 2013 by United Nations Development Program (UNDP), Indonesia has a Gender Inequality Index (GII) score of 0.494, ranking 106 out of 148 countries in 2012. The result reflects that Indonesia has loss in the equality between male and female in human development in the three GII dimensions: (1) reproductive health (maternal mortality and adolescent fertility rates); (2) empowerment (share of parliamentary seats and attainment at secondary education) and (3) economic activity (labour market).

As in Table 4.7, in the year 2012, only 51.2 % of female in Indonesia participated in the labour market as compared to 84.2 % for male. 18.2 % of female held the parliamentary seats, and 36.2 % of women have a chance to study until secondary or higher level education in Indonesia. If compared with other region in Asia, for instance, Philippines and China, Indonesia still struggles to achieve gender equality development.

All these indexes indicate that the country predominantly depends on men in most businesses. Such gaps and inequalities are prevalent mainly in the rural areas making it difficult for women entrepreneurs to carry on their business activities.

4.5.1 Research Conducted in Indonesia

Citing another research conducted by the author in Indonesia (Vong 2010), where focus group interviews were conducted with 70 respondents (comprising 45 women entrepreneurs and bank staff of which a majority are women), on bank service

Table 4.4 Indonesia gender gap index (GGI) 2012 (Hausmann et al. 2012)

Gender gap index	Overall		Economic participation		Educational attainment		Health and survival		Political empowerment	
	Rank	Score	Rank	Score	Rank	Score	Rank	Score	Rank	Score
2012 out of 135 countries	97	0.659	104	0.565	92	0.973	107	0.966	73	0.132
2011 out of 134 countries	90	0.659	101	0.564	93	0.967	106	0.966	61	0.14
2010 out of 134 countries	87	0.659	100	0.575	95	0.964	105	0.966	58	0.141
2009 out of 134 countries	92	0.658	100	0.572	95	0.966	87	0.972	70	0.122
2008 out of 130 countries	93	0.647	90	0.571	97	0.945	82	0.972	80	0.101
2007 out of 128 countries	81	0.655	82	0.599	93	0.949	81	0.972	70	0.101
2006 out of 115 countries	68	0.654	67	0.598	81	0.949	88	0.969	63	0.101

Table 4.5 Ranking by region 2012 (Social Watch 2012)

Asia and the pacific					
Country	Score	Rank	Country	Score	Rank
New Zealand	0.7805	6	Bangladesh	0.6684	86
Philippines	0.7757	8	Maldives	0.6616	95
Australia	0.7294	25	Indonesia	0.6591	97
Sri Lanka	0.7122	39	Malaysia	0.6539	100
Mongolia	0.7111	44	Japan	0.653	101
Singapore	0.6989	55	Cambodia	0.6457	103
Thailand	0.6893	65	India	0.6442	105
Vietnam	0.6867	66	Korea, Rep.	0.6356	108

Table 4.6 Gender equity index 2012 for ASEAN countries

Country	Education	Economic activity	Empowerment	GEI
Brunei Darussalam	99	78	39	72
Cambodia	71	73	21	55
Indonesia	93	57	36	62
Lao, PDR	75	69	23	56
Malaysia	98	40	31	56
Philippines	100	67	61	76
Singapore	94	71	40	68
Thailand	97	77	39	71
Viet Nam	95	75	41	70
Myanmar	97	65	*	*

*There are no available data on GEI

Source Hausmann et al. (2012), Social Watch (2012)

Table 4.7 Indonesia's GII for 2012 relative to selected countries and groups

	GII value	GII rank	Maternal mortality ratio	Adolescent fertility rate	Female seats in parliament (%)	Population with at least secondary education (%)		Labor force participation rate (%)	
						Female	Male	Female	Male
Indonesia	0.494	106	220	42.3	18.2	36.2	46.8	51.2	84.2
Philippines	0.418	77	99	46.5	22.1	65.9	63.7	49.7	79.4
China	0.213	35	37	9.1	21.3	54.8	70.4	67.7	80.1
East Asia and the Pacific	0.333	–	73	18.5	17.7	49.6	63	65.2	80.6
Medium HDI	0.457	–	121	44.7	18.2	42.1	58.8	50.5	79.9

Source United Nations Development Programme (2013)

delivery quality, core competence of bank staff and product features for women entrepreneur. The author has since re-visited the responses of the focus group interviews and discovered many subtle messages relating to constraints of women entrepreneurs in Indonesia. These messages can be viewed from the perspectives of Indonesian culture, women education, education status of women, cost of bank transactions, and mobile payments.

4.5.1.1 Indonesian Culture

The various evidences used for the hypothesis analysis are interchangeably related to such extent that it may be difficult to determine which are the cause and the ripple effects. The description of womanhood by the state government roots from the Javanese cultural values and Islamic values. The social construction Indonesian womanhood is based on three concepts: 'Istri' (housewife), 'Ibu' (Mother), and 'Ibuism', which symbolises their position in the context of a patriarchal system. Due to this, an Indonesian woman is not considered to be a bread winner and perceived as a non-productive member in the society (Ida 2001).

Indonesian country can be divided into two regions based on economic development: urban region and rural region. One of the major dissimilar characteristics would definitely point to the varying degree to which socio-cultural norms are practised in each region (Davis and Abdiyeva 2012). This paper also eloquently states that women in rural areas face certain structural, cultural, and institutional constraints. This creates difficulty to improve female educational attainment as many households in the society subscribe to the belief that education only belongs to men only.

The culture restricts women from participating in public places at mixed-gender events or groups. This reduces their opportunity to attend skill-building programmes (Kantor 2005). However, Davis and Abdiyeva (2012), states that it is imperative for micro-entrepreneurs to have the ability to secure credit and other financial requirements. It is quintessential even more so for people in less industrialized or developing economies as it requires networking which is considered a taboo in the Indonesian culture for females.

In the literature review done by Davis and Abdiyeva (2012), it is inferred that the suppressions of women by patriarchal norms and masculine culture is a common theme in women entrepreneurship, irrespective of the size of the firm. In the Asian society, socio-cultural expectation of women is a familial role, which endows them with undivided childcare responsibilities, when the men are considered to have the breadwinner role. This can lead a woman to choose to start microenterprises, in order to balance their socially inherited roles and constrain her spatial mobility, which is required to access formal credit.

4.5.1.2 Education Status of Women Entrepreneurs

The education status of a female entrepreneur is an important variable while understanding their difficulties (Huarng et al. 2012). As a whole, Indonesia is composed of 33 provinces, 98 Cities, 8,072 Urban Villages (12 %), and 67,172 Rural Villages (88 %). Indonesia's population as of 2011 is 242.3 million (The World Bank 2013). The rural population of Indonesia, as recorded by World Bank till 2010, is 46.30 % of the total population (Trading Economics 2012). With a sex ratio of 1:1, it can be said that there are approximately 56 million women living in rural areas (Central Intelligence Agency 2013).

In many cases, women are socially perceived as not needing education because the task of a married woman frequently revolves around housework and raising children. Some families also believe that it is better to educate men rather than women because married men are assumed to take social and financial responsibilities for their family as well as their parents. Hence, it seems rational for them to educate males in order for them to obtain better work opportunities. This is a social and cultural barrier with respect to the task of empowering women.

The number of women participation in business has increased every year due to the increase in number of literate women. In Indonesia, families particularly those in rural areas, still do not find the need to send their daughters to school beyond the elementary level. When Indonesia suffered an economic crisis in 1997, women and girls were the first to sacrifice their position in jobs and schools. The gap between men and women has barely changed at higher levels of education (Prasilowati 2000). As the years have passed, there has been an increase in the literacy rate of women, especially in terms of primary education. Among female population that attends primary school, 1 out of 5 rural profiles don't graduate to the junior secondary level. This is a wide gap when compared to the ratio of 1:10 in urban profiles (UNICEF 2012).

As per 2004 estimates, Indonesia's literacy rate was recorded as 94 % for males and 86.8 % for females (Vong 2010). In 2011, the literacy rate improved to 95.6 % for males but reduced to 90 % for females. The female literacy rate, especially in the rural areas, was recorded as 86.30 % in 2011 (Badan Pusat Statistik 2011).

In contrast, there are some reports, which state that the actual illiteracy rate of rural women is 2–3 times higher than the reported government data. It also states that many of them cannot speak their own language i.e. Bahasa Indonesia. In spite of the efforts made by the government to eliminate illiteracy, they seem to be ineffective due to lack of commitment of the Head of the Region.

Government spending on literacy has witnessed a steady and commendable increase in Indonesia. Indonesia invested around $13.1 billion in the educational sector in 2006, which increased to $16.5 billion in 2009. Educational expenditure as a percentage of GDP was 3.16 % in 2009 (Vong 2010). As one of the Millennium Development Goals, special attention is now given to Women Empowerment in Indonesia. They have realized that education plays a vital role for accomplishing this goal. Government has now allocated 20 % of the national budget to increase the number of children receiving junior secondary education. To guarantee the right to education for every girl child, it has also taken necessary measures to ensure 9 years of basic education for every Indonesians child. Furthermore, government provides formal education packages for children with special needs and informal education for out-of-school young women.

According to OECD report (Hung et al. 2012), there has been good progress in terms of education to women in primary, second and tertiary education. However, in spite of this, economic empowerment of women is far lesser than men. Although the law does not tolerate discrimination, cultural influences that discourage women to be full-time entrepreneurs still prevail in many rural areas.

4.5.1.3 Financial Education Among Females

The research work in Indonesia and India highlights that financial literacy has a strong influence on demand for financial services (Cole et al. 2009). According to OECD definition, financial literacy is a combination of cognitive attributes necessary to make sound financial decisions and ultimately achieve individual financial well-being (Hung et al. 2012).

A wide range of empirical literature states that gender differences do exist in financial literacy along various dimensions. On average, the performance of women on tests of financial knowledge in different countries has proven to be worse than men, and have less confidence in their financial skills. Many literatures have identified wide-spread potential repercussions of gender differences in many countries. Lower levels of financial literacy may possibly lead to reduction in women's active involvement within a country's economy (Hung et al. 2012).

It is noticed that, when women have access to information on market opportunities and financial services that are available to them, they may not be well equipped to avail it. Another very important question is whether there is any difference between men and women in their willingness to take risks. Studies in economics and psychology have identified that women, on average, tend to be more risk averse than men. Moreover, they may even give up activities that require them to bear excessive risk although these opportunities might offer higher returns. Women are more likely to consider borrowing against collateral as a risky transaction and might be less interested in taking out loans even when credit is available to them (Fletschner and Kenney 2011).

A report by Fletschner and Kenney (2011) inferred that women, as compared with men, follow social norms with a different set of family rights and responsibilities, tend to have limited access to information, less control over resources that are accepted as collateral, and more risk averse. It can be clearly understood that, the conditions under which women are ready to participate in formal financial markets, the type of financial products and services they require, and their ability to fulfil their financial needs varies considerably from the requirements of men. These findings suggest that women will have a stronger preference for financial products that are customized and personalized to allow them to borrow without any risk of losing their assets, to save in a safe and secured environment, and also insure against any risks (Fletschner and Kenney 2011).

4.5.1.4 Cost of Bank Transactions

One reason for poor women to participate in informal financial mechanism is transaction cost. According to Vonderlack and Schreiner (2002), they found that minimum transaction costs are required for women to use financial services. For those poor women opportunity cost of time and transaction costs, such as childcare fee, transportation, copies needed to open an account can overburden all other factors in the options of savings mechanism (Vonderlack and Schreiner 2002). Therefore, it is recommended that financial institutions should offer a one stop platform, "one system, one clearing house and one day clearing" for cheque transactions (Bank of Thailand 2008). In this platform, micro women entrepreneurs can reduce time and improve convenience, instead of handling many passbooks and visiting many banks to transfer money. Technically, it is called as Imaged Cheque Clearing System (ICS) and Image Archive System (IAS). It has integrated Electronic Cheque Clearing (ECS), Provincial Cheque Clearing, and Bill for Collection (B/C) into a single system (Bank of Thailand 2008). As the result, the cheque processing is reduced from 2 or 3 working days to 1 day, to save on transaction and opportunity costs and to meet the women's entrepreneur needs.

Mobile phone companies and network providers play a vital role in promoting mobile banking among its customers. One such example is Simpay, a consortium formed by several European Telecom service providers. Its main aim was to offer a single common platform to process routing, clearing and settling of m-payments. However, in 2005, Simpay collapsed even before it started its operations due to several impediments. Its biggest challenge was to bring rival firms together and provide equal benefits to everyone. Hence, it is argued by many that m-payment needs cross-industry involvement to be successful. For instance in Japan, NTT DoCoMo has aggressively teamed up with many financial institutions and has spent ¥99 billion on technology to implement m-payment. They achieved success, as within a span of less than 1 year about 25000 retail outlets started accepting m-payments from customers. There have been many such success stories in South Korea as well. Hence, it is said that a joint venture between mobile network operators and banks, aided by technology producers, is needed to address current challenges related to product development, information security, users' requirements, resource and expertise sharing, and so forth (Au and Kauffman 2008).

4.6 Recommendations

4.6.1 Micro Financial Products Design

Waterfield (2001) reinforces that credit institutions require incorporating financial products and delivery methodologies, which are both appropriate and sustainable. The author insists that micro financial institutions may have to be three hundred times more efficient than commercial bank during loan disbursement activities. This is vital as microfinance products have tendencies to deviate from the majority of products employed by formal lending institutes.

There are various limitations in the ability of formal and informal sources of finance. These limitations can exaggerate further in reference to the women micro-entrepreneurs in Indonesia. Table 4.8 shows two types of sources of microfinance products. Microfinance products would require relaxing the collateral criteria and developing more evaluative tools to analyse loans and character of the women micro-entrepreneur. Secondly, it is necessary to devise a system to prepare alternative plans if micro-loans were to default but nonetheless minimize the administration expenses during the process (Waterfield 2001).

Waterfield (2001) exemplified the successful loan repayment rates when poor women micro-entrepreneurs involved in informal source schemes such as ROSCA's, where peers in the group place pressure on each other in order for the whole group to be credible for the following loans. The informal source from moneylenders brings to notice that formal micro financers are required to make quick response to needs of women micro-entrepreneur in a timely and non-bureaucratic fashion.

Most banks offer a good range of micro financial products and services. However, they still lack effort in attracting women entrepreneurs from rural areas. Focus areas for banks and financial institutions in relation with women entrepreneurs in Indonesia:

Table 4.8 Sources of microfinance products

Informal sources		Formal sources
Rotating savings and credit associations (ROSCAs)	Moneylenders/middlemen	Banks and coops
Inflexible timings of loans	Extremely high interest rates	Very limited access
Limited and inflexible loan amounts	Often limited loan amounts	High transaction costs
Risk of loss of investment	Obligations to sell to middle-man at sub-market rates or buy expensive inputs	Rigid collateral requirements

4.6.1.1 Reduce Informal Finance Dependency

In case of micro financing, banks charge high interest rates, have rigid and lengthy procedures, and require collaterals. On the other hand, although interest rates of informal finance are high, they are flexible in terms of collaterals and efficient with loan approvals (Riedinger 1994). Banks can counter high interest rates by reducing operational cost. This can be done by eliminating middlemen, distribution agents, physical office buildings, etc. and make use of technologies like biometrics, mobile banking, mobile wallets, etc. Mobile banking reduces operational cost by almost 50 % (Columbia Business School 2012). They need to create a women friendly environment by creating an aesthetic appeal, personalizing conversations, and fulfilling their unstated needs in an efficient manner. According to a pilot study conducted in Cambodia, it was observed that the impact on introduction of mobile banking has created social inclusion among the rural population in four major dimensions: consumption behaviour, participation in economy, political contribution and social network creations (Vong et al. 2012).

4.6.1.2 Create Awareness of Financial Services

Rural women, as mentioned earlier, have less access to secondary education and also display poor financial knowledge. Hence, MFI's and banks need to conduct awareness programs in every rural area on a regular basis. Currently, the most common services that customers would like to use on a daily basis are—transactions between bank account holders, top-ups, remittance, repayment, buying things with micro-amounts (International Finance Corporation 2010a). All these services can be easily accessed with the help of mobile banking. Hence, the benefits of mobile banking and its success stories in other countries like Kenya should be highlighted to the masses in rural areas. In order to make this program more effective, financial institutions need to collaborate with popular telecom services like Indosat and Telkomsel. For the payment and banking transactions, customers can use only the cellular network; no Wi-Fi is needed. This will provide access to majority of the rural women entrepreneurs as there are about 180 million mobile subscribers in Indonesia (Vong 2010). They can also introduce mobile vans, which will not only provide basic financial assistance to remote areas, but will also work as an advertising mechanism.

4.6.1.3 Use of Technology to Lower Bank Transaction Cost

Information and Communication Technologies (ICT) becomes a vital dynamic tool for entrepreneurial, firms and social opportunities and human development. ICT is "prime mover" that rapidly decreases social inequality (Chen and Wellman 2005) and has significant role in economic development and poverty reduction (African Partnership Forum Proceedings 2008).

As women represent the major economic driven in many countries, the use of ICT, information-based or knowledge-based in women is growing for countries' economic as well as improving quality of life to both of urban and rural regions.

In 1995, The Fourth World Conference on Women in Beijing explained that the utilizing of mobilization, information exchange, and empowerment in women is potential tool for understand about information technology more and more. Unfortunately, women in some areas are likely to have barriers to access ICT, including the Internet, for example language problem, low levels of literacy rate computer illiteracy and insufficient telecommunications infrastructure. "Technophobia" is one of constrains for women attitude towards ICT. Furthermore, in less developed countries like Africa, patriarchal gender norms are obstacles for women's connectivity to the Internet. Thus, the gender limitation and low levels of literacy rate can results to micro women entrepreneur's access to ICT.

However, Web 2.0 (Social networking), mobile banking, handheld personal digital assistants (PDAs) and fingerprint IDs can now offer those groups to reach formal micro financial services in real-time (Kauffman and Riggins 2012). As the result, bridging improved comprehension for micro women entrepreneurs to digital devices can increase opportunity micro financial operations.

Hence, in a framework called microfinance technology and stakeholder ecosystem, it is considered that Information and Communication Technology (ICT) will play an important role to make microfinance more effective in its outreach. There are some challenges that Microfinance Providers (MFP) currently face in terms of ICT—lack of investment in technology, inadequate internal IT capabilities, and security issues related to execution of financial transactions (Kauffman and Riggins 2010).

4.6.1.4 Value Added Services

Organizing conferences, networking seminars and business education sharing workshops will provide an excellent platform for women entrepreneurs, especially for those who have newly established their ventures. This will also allow successful entrepreneurs to share their experiences and best practices. These types of additional services will need to be an on-going process. Apart from financial literacy of rural women, these seminars will need to cover various topics such as specific health related issues, resource and financial management.

4.6.1.5 Use of Mobile Banking

In the last decade, technological advancement has gained attention due to the addition of mobile (m)-payments. A study done by Royal Philips Electronics and Visa International shows that consumers like to use mobile phones for transactions and payments, as it is convenient and easy to use. In addition, a research-consulting firm called Celent states that in 2006 approximately $24 billion payments were made through mobile phones, which significantly increased to $55 billion in 2008.

Furthermore, a survey conducted by Visa USA among 800 Americans revealed that over 50 % of the people surveyed between age group 18 and 44, preferred m-payments over cash for small purchases.

Most of them also stated that they would like to carry just mobile phones instead of physical money. Similar results were found in London as well. According to analysts and researches mobile payments will continue to prosper as more people start to understand the benefits of technology and digital wireless phones (Au and Kauffman 2008).

Another mobile banking system designed for microfinance is ACMB i.e. Affective Cashless Mobile Banking. Its prime objective is to maximize the outreach of banking services to the vast unbanked population in a cost-effective and usable manner. It also helps to secure mobile banking transactions by implementing credit card transaction protocol. ACMB's test and survey showed promising results and positive responses from users as most of them found it effective, efficient and easy to use. For instance, creating a new account took less than 3 min and the system could handle 804 transactions in 3 h. However, to make it an affordable banking system, it needs to operate on 2G cell phone networks, as 3G networks are not yet implemented in many developing countries. This could lower the cost of owning a banking system to US$100, which will be capable of supporting about 2000 transactions (Song and Vong 2013).

4.6.2 Micro-insurance

In patriarchal system of Indonesia, women may be described as "low-asset people". Therefore, the provision of micro-insurance along with micro-loans act as the protection against specific perils in exchange for regular premium payments proportionate to the likelihood and cost of the risk involved. Micro insurance products can be distinguished into those with a savings element and those that are pure insurances. The former product is more popular as it has a balanced reciprocity arrangement. Micro insurance is essentially seen as a safety net to maintain the status quo when it is connected to microcredit and micro savings. The existence of a micro-insurance reduces the applicant's risk, which will boost their confidence to continue use of micro-credit and also allow for easier access in the eyes of formal micro financiers.

With inspiration from two models devised by Hintz (2009), a provider driven-community based in which multiple but small group of applicants, in this case rural women micro entrepreneurs, on formal terms gather their resources and risks into one pool. It is said to be provider driven, as the insurance provider will run the entire micro insurance scheme including its distribution. This would require the establishment of female insurance agent network who would educate and interact with the client groups. Micro insurance is the new tool to overcome collateral security issues of women entrepreneurs. Banks can either offer it themselves or can guide customers to such insurance companies. It would be best for banks to handle micro insurance process so that customers can deal with only one institution for all services.

4.7 Conclusion

This paper has shown that women micro entrepreneurs face multiple forms of constraints, but they originate from an underlying gender inequality in their country of origin. With reason to believe that altering socio-cultural taboos is high on the government's agenda, the process will be time consuming and the rate of change will be comparatively slow. While working around this shortcoming, the opportunity for women micro entrepreneurs can be addressed by micro financiers. However, the gender inequality issues should be their focus of attention during the design process of successful micro-financial products. Therefore, there is potential for further research possibilities to conduct field tests in rural Indonesia to explore the hypothesis generated. This effort will empower the entrepreneurial success of a developing country and help to alleviate poverty levels, irrespective of the various coercions from within the society.

References

African Partnership Forum Proceedings (2008) In: 2008 Partnership forum for making finance work for Africa Accra, Ghana
Au YA, Kauffman RJ (2008) The economics of mobile payments: understanding stakeholder issues for an emerging financial technology application. Electron Commer Res Appl 7(2):141–164
Bank of Thailand (2008) Imaged cheque clearing and archive system. http://www.bot.or.th/English/PaymentSystems/PSServices/ChequeClearingSys/ICAS/Pages/ImagedCheque.aspx, 2013
Central Intelligence Agency (2013) The world factbook. https://www.cia.gov/library/publications/the-world-factbook/geos/id.html, 2013
Chen W, Wellman B (2005) Minding the cyber-gap: the internet and social inequality. In: The Blackwell companion to social inequalities. Blackwell, Malden, pp 523–545
Cole SA, Sampson TA, Zia BH (2009) Financial literacy, financial decisions, and the demand for financial services: evidence from India and Indonesia. Harvard Business School, Boston
Columbia Business School (2012) Microfinance innovation: technology platforms enabling MFIs to broaden impact. Online media (2014). http://www.youtube.com/watch?v=g-Ft1-PZLTU
Davis PJ, Abdiyeva F (2012) En route to a typology of the female entrepreneur? Similarities and differences among self-employed women. J Manage Policy Pract 13(4):121–137
Eversole R (2003) Help, risk and deceit: microentrepreneurs talk about microfinance. J Int Dev 15 (2):179–188
Fletschner D, Kenney L (2011) Rural women's access to financial services'. ESA working paper no. 11
Hausmann R, Tyson LD, Zahidi S (2012) The global gender gap report. World economic forum
Hintz M (2009) Micro-Impact
Huarng K-H, Mas-Tur A, Yu TH-K (2012) Factors affecting the success of women entrepreneurs. Int Entrepreneurship Manage J 8(4):487–497
Hung A, Yoong J, Brown E (2012) Empowering women through financial awareness and education. OECD Publishing: http://www.finlitedu.org/team-downloads/evaluation/hung-yoong-brown-2012pdf.pdf
Ida R (2001) The construction of gender identity in Indonesia: between cultural norms, economic implications, and state formation. Surabaya: Jurnal Masyarakat, Kebudayaan, dan Politik 14 (1):21–34

International Finance Corporation (2010a) Mobile banking in Indonesia. International Finance Corporation, Washington DC
International Finance Corporation (2010b) Serving the financial needs of Indonesian SMEs: Main findings. World bank report. International Finance Corporation, Bethesda
International Finance Corporation (2013) Banking on women. http://www1.ifc.org/wps/wcm/connect/Industry_EXT_Content/IFC_External_Corporate_Site/Industries/Financial+Markets/MSME+Finance/Banking+on+Women/
Johar M, Rammohan A, Economics (2006) Demand for microcredit by Indonesian women. University of Sydney, Sydney
Kantor P (2005) Determinants of women's microenterprise success in Ahmedabad, India: empowerment and economics. Feminist Econ 11(3):63–83
Kauffman RJ, Riggins FJ (2010) Research directions on the role and impact of ICT in microfinance. In: 43rd Hawaii international conference on system sciences (HICSS), 2010. IEEE, pp 1–10
Kauffman RJ, Riggins FJ (2012) Information and communication technology and the sustainability of microfinance. Electron Commer Res Appl 11(5):450–468
Kuehn D (2012) The financial challenge facing women. your money. American Federation of Teachers, Washington DC
Narain S (2009) Gender and access to finance. World Bank Group, Washington DC
Prasilowati SL (2000) An analysis of women's education in Indonesia: empowerment and barriers, Masters Theses, Patrick Power Library, Saint Mary's University, Halifax, Canada
Riedinger JM (1994) Innovation in rural finance: Indonesia's Badan Kredit Kecamatan program. World Dev 22(3):301–313
Roper S, Scott JM (2009) Perceived financial barriers and the start-up decision an econometric analysis of gender differences using GEM data. Int Small Bus J 27(2):149–171
Singh SP, Comer S, Catlin CJ, Reynolds RG, Sutanto A (1999) A profile of microenterprises in rural Indonesia. http://www.sbaer.uca.edu/research/icsb/1999/110.pdf, 2013
Social Watch (2012) Gender equity index by country. Social Watch
Song I, Vong J (2013) Affective core-banking services for microfinance. In: Computer and information science. Springer, pp 91–102
Statistik BP (2011) Statistics Indonesia. http://www.bps.go.id/eng/tab_sub/view.php?kat=1&tabel=1&daftar=1&id_subyek=40&notab=9, 2013
Tambunan T (2009) Women entrepreneurship in Asian developing countries: their development and main constraints. J Dev Agri Econ 1(2):27–40
Tambunan TTH (2011) Development of small and medium enterprises in a developing country: the Indonesian case. J Enterprising Communities People Places Global Econ 5(1):68–82
The World Bank (2013) Indonesia—data. http://data.worldbank.org/country/indonesia, 2013
Trading Economics (2012) Rural population in Indonesia. http://www.tradingeconomics.com/indonesia/rural-population-percent-of-total-population-wb-data.html, 2013
UNICEF (2012) 2011 Annual reports. http://www.unicef.org/about/annualreport/index_62542.html. 2013
United Nations Development Programme (2013) Human development report 2013. United Nations Development Programme, USA
Vial V (2011) Micro-entrepreneurship in a hostile environment: evidence from Indonesia. Bull Indonesian Econ Stud 47(2):233–262
Vonderlack RM, Schreiner M (2002) Women, microfinance, and savings: lessons and proposals. Dev Pract 12(5):602–612
Vong J (ed) (2010) Women banking strategy for bank international Indonesia. IFC World Bank, Washington DC
Vong J, Fang J, Insu S (2012) Delivering financial services through mobile phone technology: a pilot study on impact of mobile money service on micro–entrepreneurs in rural Cambodia. Int J Inf Syst Change Manage 6(2):177–186
Waterfield C (2001) Designing microfinance loan products. MFI Solutions, LLC :1–16

Chapter 5
Lowering the Interest Burden for Microfinance

Abstract MFIs have a high interest rate burden due to the small monetary amount per transaction of microcredit and an inevitably high operating cost per transaction. To ensure financial viability and to expand the depth and breadth of their operations, MFIs have to adopt cost recovery interest rates on microcredit, hence MFIs have to charge interest rate high enough, usually substantially higher than the bank loan risk free interest rate. The major factors in determining the interest rate on microcredit are the cost of funds, operating costs, loan loss cost and capital for business expansion. To illustrate the impacts of the above factors on interest rate, we present a summary of the current cost structures of MFIs in three Southeast Asia countries, Cambodia, Vietnam, and Indonesia. Then, we shall review existing studies and propose new uses of mobile technologies and financial market innovations for lowering interest burden.

Keywords Microfinance · Mobile technologies · Insurance · Interest burden

5.1 Introduction

Access to reliable and affordable financial services such as savings, credit, payments, transfers, and insurance are vital to living in a modern economy (Tan and Teo 2000), yet over half of the world population are still unbanked. Chaia et al. (2010), CGAP (2010) and Ardic et al. (2011) suggested that as of the end of 2009, 2.75 billion people (56 % of the global population) do not have access to formal financial services. A recent survey of 150,000 adults aged 15 and above in 148 different economies conducted by World Bank reported that over 50 % of the respondents do not have any account in any financial institutions. While 89 % of adults in developed countries have access to financial services the percentage is only 41 % in developing countries.

For the unbanked, poverty (no cash on hand), relatively high costs associated with accessing banking services, and inconvenient locations (long travelling

J. Vong and I. Song, *Emerging Technologies for Emerging Markets*,
Topics in Intelligent Engineering and Informatics 11,
DOI 10.1007/978-981-287-347-7_5

distance) of branches are reported as the major reasons for not using any banking services. A third (approximately 800 million people) of the unbanked population in Asia, who are in the lowest income category (i.e. living on under $5/day), simply cannot afford the high costs of commercial banking services. For commercial banks, providing banking service in these low-income communities is difficult. The main reasons are high fixed costs of banking systems and the costs for running banking activities (e.g., credit scoring and lending) required by commercial banks. Some commercial banks also hesitate to participate or invest in these markets due to lack of confidence and to avoid supporting potential future competitors. The mechanism to provide a self-sustainable business model to serve these markets continues to be a challenge.

In the last few decades, many MFIs have emerged to provide financial services and microcredit to the unbanked sector. For example, solidarity lending and village banking pioneered by Grameen Bank (Yunus 1999) in Bangladesh, Faulu M-PESA (Hughes and Lonie 2007) and Kenya Women Finance Trust (KWFT) (www.kwft.org) (Kiiru and Pederson 1996) in Kenya, WING in Cambodia (www.wingmoney.com) (WING 2009), and many others. It is believed that microfinance is an effective instrument to alleviate poverty by providing financial service, particularly microcredit and savings accounts to the low-income individuals and microenterprises (Hammond et al. 2007; World Bank 1999).

Although microfinance began in Bangladesh in the 1970s, after 30 years, there is still half of the world's population outside the reach of MFIs. The majority of existing microfinance solutions are operated on labour-intensive business models that involve posting transactions from one ledger to another with human hands. These not only pose a substantial challenge for these MFIs to scale fast enough to serve the huge demand from the unbanked population, but also create difficulties to be accounted for illegitimate activities and frauds. In particular, the substantial portion of operating costs is due to manual labour. Credit officers have to maintain frequent contact and close relationship with customers to enforce loan contracts and to control risk of lending as the MFI customers cannot provide collateral or credit histories. Operating costs are inevitably high for each loan transaction as well. Therefore, MFIs charge much higher interest rates than banks to ensure the permanence and expansion of the services without ongoing needs of subsidies from donations. Therefore, the loan interest rates of MFIs range from 17 % (in Sri Lanka) to 80 % (in Uzbekistan), with a global average of 35 % CGAP (2008). The high microcredit interest rates have been heavily criticized and a number of countries have tried to legislate interest rate cap on microcredit. In summary, for MFIs to scale fast enough to serve the huge demand from unbanked population and maintain financial sustainability, they require new business models and innovations to lower the interest burden by (1) lowering the operating cost, (2) accounting for legitimate activities and frauds, (3) lowering or mitigating risks of microcredits.

This study aims to understand the causes of interest burden of MFIs and discusses the roles of mobile technologies and financial market innovations for MFIs to lower the interest burden. To illustrate the impacts of the various factors on interest rate, we present a survey of current cost structures of microfinance institutes (MFIs) in Southeast Asia. Then, we review existing and propose new uses of mobile technologies and financial market innovations for lowering the interest burden.

5.2 Impact of Costs on Interest Burden

For MFIs to be financial viable, they need to use their loan interest income to cover the operating costs which comprise of cost of funds, loan loss expenses, and other operating expenses (e.g., equipment and wages) (Rosenberg et al. 2010):

$$\text{Income from loans} > \text{Cost of funds} + \text{Loan Loss expenses} + \text{Operating expenses} \quad (5.1)$$

Lowering interest burden would require lowering one of the three components on the right side of the equation. In this section, we will discuss how these components may affect interest income of an MFI. Financial information on MFIs is drawn from the database of the Microfinance Information Exchange (MIX). Not all MFI report to MIX, but those that currently do are over 2,000 MFIs worldwide in 110 countries. We use the adjusted financial information of these MFIs to compensate for the effect of any subsidies they receive and, thus, try to present a picture of what the industry would look like if they had to pay market costs for all of their resources. We present the cost structure of MFIs in three Southeast Asia countries: Cambodia, Vietnam and Indonesia. We collected a dataset that includes 58 MFIs that report their results to MIX in 2011. 9 out of the 58 MFIs have a negative net income after taxes and before donations; hence they are not financially viable. They are excluded from this analysis. Among the 49 financial viable MFIs, there are 15 from Cambodia, 21 from Vietnam, and 14 from Indonesia.

Looking at interest rates, we use percentage of income from loan on gross loan portfolio (GLP), which is the total amount that cash borrowers pay the MFI during a period for interest and fee from loan divided by the average outstanding GLP over the same period. Figure 5.1 shows the minimum, 25th percentiles, medians, 75th percentiles and maximum interest and fee incomes of MFIs as % of GLP. The values range from 16.7 to 37.97 % in Cambodia, 10.33 to 28.74 % in Vietnam, and 17.77 to 83.09 % in Indonesia. During the same period, the average risk free interest rates in these countries are 1, 9 and 6 %.

To investigate the factors related to the interest rate of MFIs, we look at the different costs and taxes of MFIs as percentage of income from loan. Figure 5.2 shows the minimum, 25th percentiles, medians, 75th percentiles and maximum (a) cost of funds, (b) loan loss cost, (c) operating cost, (d) profit and (e) taxes of MFIs as % of total income from loan during 2011.

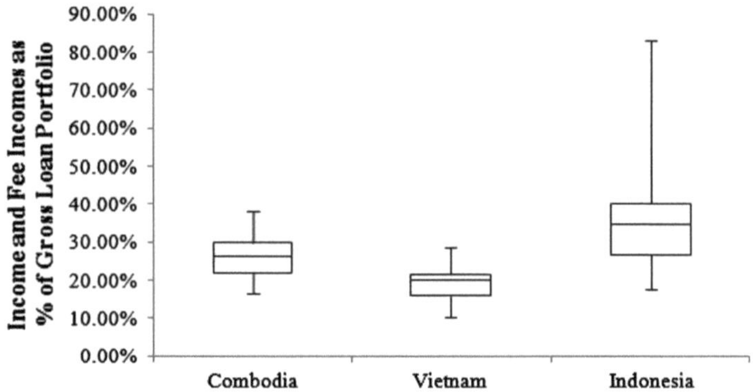

Fig. 5.1 Interest income as percentage of gross loan portfolio for 2011

The costs of funds as percentage of income from loan for the MFIs are relatively high with medians of 25.43 % in Cambodia, 13.4 % in Vietnam and 27.37 % in Indonesia. The high cost of funds as percentage of income from loan may be due to MFIs paying more expense interest rate to borrow capitals to fund their loan portfolio; however, as compared to commercial banks, MFIs tend to have less of their loan portfolio funded by liabilities. Another possible reason for this may due to poor liquidity management for MFIs, MFIs may tend to keep more cash in hand for liquidity risks.

Loan losses due to microloan default have relatively little effect on MFI interest burden, loan loss as percentage of income from loan for MFIs have medians of 0.01 % in Cambodia, 0.76 % in Vietnam and 1.17 % in Indonesia. MFIs usually have default rate well below those of commercial banks in their countries. MFIs with too low loan loss rate may be too risk-averse in their selection of borrowers, which may dampen the reach of MFIs to the poor.

Substantial portion of MFIs income from loan goes to pay operating costs. This becomes the major factor of high interest burden for MFIs. Operating cost as percentage of income from loan for MFIs has medians of 49.28 % in Cambodia, 53.28 % in Vietnam and 47.21 % in Indonesia. There are a large range of factors that affect operating costs of MFIs, including loan size, scale of MFIs, client location and density, communication and transport cost, type of loans, office and equipment costs, and salary levels. Lower operating cost of MFIs will lower the interest burden of MFIs considerably.

Profit has a mild effect on interest burden on MFIs. Profit as percentage of income from loan for MFIs has medians of 17.06 % in Cambodia, 32.67 % in Vietnam and 19.61 % in Indonesia. Making profit from microfinance is controversial, as MFIs are viewed as a service to the poor, not as a profit making opportunity. However, one may argue that profit of MFIs can attract more investment in MFIs and allow more rapid outreach of MFIs to the poor with the capital to expand their operations.

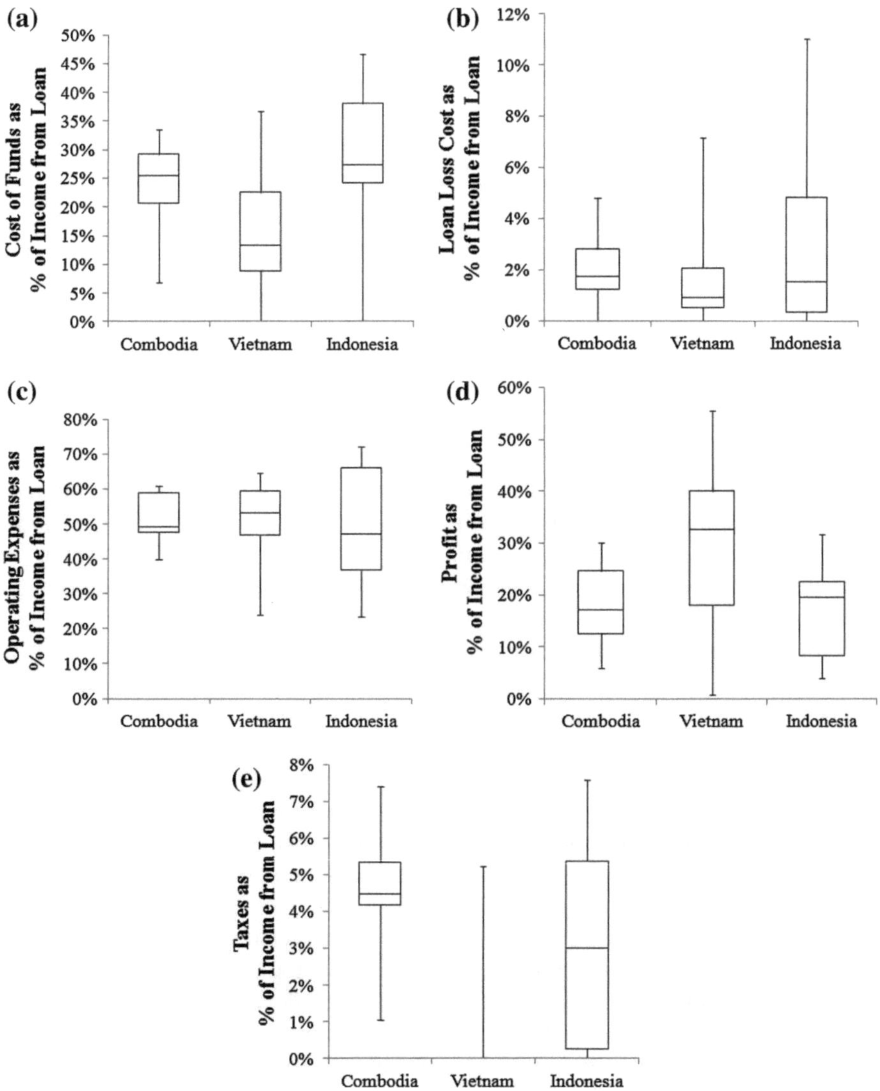

Fig. 5.2 **a** Costs of funds of MFIs as percentage of income from loan for 2011. **b** Loan loss costs of MFIs as percentage of income from loan for 2011. **c** Operating expenses of MFIs as percentage of income from loan for 2011. **d** Profit of MFIs as percentage of income from loan for 2011. **e** Taxes of MFIs as percentage of income from loan for 2011

Taxes have very little effect on interest burden on MFIs. Taxes as percentage of income from loan for MFIs have medians of 4.47 % in Cambodia, 0.00 % in Vietnam and 2.99 % in Indonesia.

In summary, to lower the interest burden of MFIs, efforts should be focus on lowering the cost of funds, lowering the operating cost and lowering the capital required for business expansion for MFIs.

5.3 Mobile Technologies Innovations

5.3.1 Mobile Technologies for Microfinance

Innovations can bring significant changes in the ways people participate in banking services. In the past decade, the use of mobile phones to provide financial services across developing countries has been one of the most remarkable technology stories. The International Telecommunications Union (ITU) reported that global mobile phone subscriptions have accelerated from 2.2 billion in 2005 to over 5 billion in 2011 reaching 86 % of the total global population and 79 % of the total developing world population. This phenomenon is set to grow to 6.8 billion by end 2013 (International Telecommunications Union (ITU) 2011). Mobile banking is believed to be one of the important drivers to enable large scale and sustainable microfinance for the poor (Kauffman and Riggins 2012).

Various terms have been introduced in the past for mobile banking, such as "Mobile Phone Banking" or "Mobile Money Services" (Beshouri et al. 2010; Donner 2008). Examples mobile banking systems include Sybase mCommerce 365 by SAP (Edgar 2009) and DELL Mobile Banking and Payments (DELL). Many studies based on mobile phone-based banking have recently been done for developed countries with regards to accessibility for people with disabilities (Pous et al. 2012), security and trust issues (Zhou 2012), public perception and attitude (Kim and Kang 2012). However, most of the existing banking systems supporting mobile phone-based banking, such as Sybase mCommerce 365 by SAP (Edgar 2009), DELL Mobile Banking and Payments (DELL) are targeted at developed countries and therefore would incur additional overheads on top of the existing banking systems. By adopting these off-the-shelf solutions or developing their own, MFIs will need to have adequate bank-office information systems. In many cases, deploying the technical solution is time consuming and expensive. For example, Opportunity Bank and XacBank, a microfinance bank in Mongolia, invested at least US$100,000 to upgrade their core banking system, before developing its own m-banking service (Kumar et al. 2010). While most people believe that mobile technologies could be used to reach more low-income customers at a lower cost, yet in the majority of developing countries, there is not yet a self-sustainable, low cost mobile banking solution that micro-banks can fully leverage.

5.3.2 *New Mobile Micro-banking Systems (MMB)*

To fully realize the benefit of mobile technology for MFIs, we developed a new mobile micro-banking model (MMB), which is (1) cashless, (2) branchless, (3) wireless, and (4) capital-less (i.e., minimum capital requirement). MMB aims to dramatically reduce the operating costs and capital requirement of individual MFIs. The target users are people in resource poor environments, such as villagers in remote areas, farmers, and unemployed who earn less than US$100 a month. The objective is to create self-sustainable micro economy within the vicinity of their residence by providing banking services, such as loan, saving accounts, and trading.

Figure 5.3 shows the operational model and economy model of MMB. Customers (farmers and house wives) receive loans (credits) from mobile operators who visit the customers to collect personal details, such as national IDs and addresses. Customers then use the credits to buy supplies for their small businesses and sell their products to other customers. All the transactions are done using mobile phones; no cash is involved at any stages. The core banking systems running on Android tablets (A and B in Fig. 5.3) automatically clear the transactions. The loans and credits are covered by the micro-insurance company who pools the

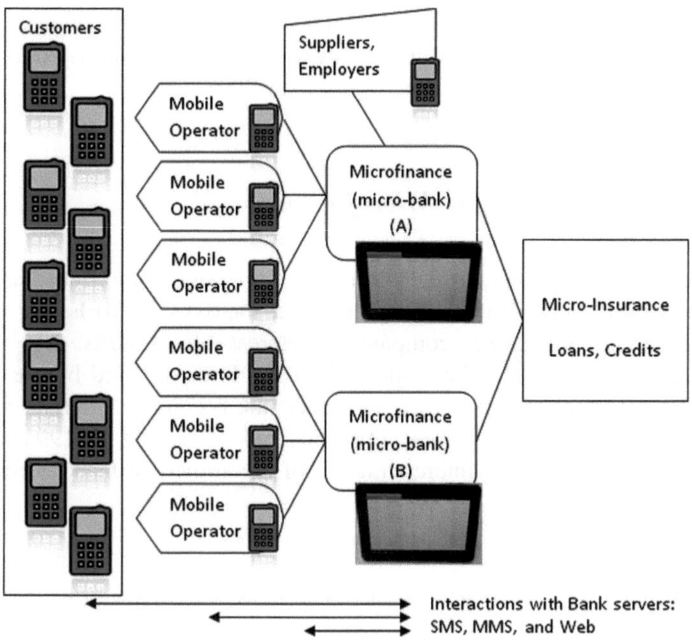

Fig. 5.3 System diagram of cashless, branchless and wireless mobile phone-based banking model. The system shows a core banking system, called MicroBanq, running on android based tablets. Bank tellers now operate on mobile offices, such as motor bikes. The insurance company pulls the risks to reduce the overall credit risk through diversification, providing low lending interest rates

risks and distribute the risks by means of insurance premiums. Technical efficiency and usability of MMB mobile banking systems have been validated (Song and Vong 2013). The total cost of ownership of the banking system could go as low as US$100. The system would be able to support about 2,000 transactions per day (assuming 10 s per transaction limited by the current SMS services).

5.4 Financial Market Innovations

5.4.1 Financial Market Innovations for Microfinance

In addition to the technological innovations, financial market innovations have important influence to MFIs interest burden. Two existing financial market innovations: (1) group lending and (2) focus lending to women are used to lower operating cost and mitigate risk in this market (Yunus 1999). These are evidence in the lower loan loss cost in most MFIs. Group lending relies on group pressure to encourage financial responsibilities of borrowers to reduce risk. Group lending also lower the operating cost of MFIs with credit officers visit and collect small weekly repayment from the group as a whole instead of from each individual. Focus lending to women not only empower women in developing countries but also results in a lower likelihood that the money will be used for non-productive purposes, hence a higher repayment rate (Garikipati 2008).

5.4.2 Micro-insurance for Microfinance

In the MMB banking model, a micro-insurance company is introduced as financial market innovation for MFIs to pool and distribute the risks for MFIs. We now show the impacts of micro-insurance company on interest rate of MFIs.

The loan interest rate R_c of a commercial bank is determined by the operating cost op_c, the risk free interest rate r_f, and other risk premiums, such as credit risk, market risk r_m, liquidity risk r_l, and operating risk (Fraser et al. 1995). For simplicity, let us define the loan interest rate R_c of a commercial bank as follows:

$$R_c = r_f + r_m + r_l + op_c \tag{5.2}$$

The transaction cost (*TrxCost*) for a micro-bank (e.g., a microfinance institute) is:

$$TrxCost = SC \times SN \tag{5.3}$$

where *SC* is the cost of sending one SMS message and *SN* is the number of SMS messages that are sent to complete one transaction. *SN* for our protocol is 2 and *SC* is usually less than US$0.01 in developing countries like Indonesia. Therefore, *TrxCost* = US$0.02.

The loan interest rate (R_{mfi}) of a MMB MFI is determined by the risk free interest rate, operation costs shared by customers, and the insurance premium:

$$R_{\mathrm{mfi}} = r_{\mathrm{f}} + op_{\mathrm{mfi}}/N + ip(L) \tag{5.4}$$

where r_{f} is the annual risk free rate, op_{mfi} is the operating cost per year, N is the number of loans (customers), and $ip(L)$ is the micro-insurance premium for loan amount L. The operating cost, op_{mfi}, of a MFI comprises of the costs of mobile operators, who visit micro-entrepreneurs and farmers to collect personal details, the banking system costs, and human resource costs for managing the micro-bank:

$$op_{\mathrm{mfi}} = mop \times VN + cb/5 + hr \tag{5.5}$$

where *mop* is the mobile operator cost per visit and *VN* is the number of visits by the mobile operator, *cb* is the core-banking system cost for 5 years and *hr* is the human resource cost. A micro–bank with around 15,000 customers requires one IT engineer and three managers (e.g., Dian Mandiri Foundation in Indonesia). It would cost about US$1,000 per an employee in Indonesia: *hr* = $4,000. Dian Mandiri handles about 2,000 transactions a day. Therefore, *cb* is US$200 and *cb*/5 = 200/5 = $40.

Now, let us suppose that the micro-insurance company has enough customers (around 10 million), then $ip(L)$ will be as low as 1 %. This means that if N is sufficiently large (e.g., 15,000), R_{mfi} converges to r_{f} + 1 %, which is the theoretical minimum interest rate of MMB micro-banking. That is, R_{mfi} is determined by the risk free rate of the macro-economy and the micro-insurance premium, not by the operating costs, credit, liquidity, and market risks.

$$R_{\mathrm{mfi}} = r_{\mathrm{f}} + 1\% = [R_c - (r_m + r_l)] + ip(L) \tag{5.6}$$

This is possible due to two reasons: (1) low operating cost per customer and (2) not-for-profit operations of microfinance (e.g., microfinance institutions cannot and do not lend money to corporations for profit). Figure 5.4 shows that the minimum number of loans that are required to self-sustainable a micro-bank.

5.5 Design and Implementation of the Mobile Banking System

The mobile banking system shown in Fig. 5.3 was developed based on the new micro-banking model and evaluated on an artificial village for its usability.

The mobile banking system defines and exports its services through Microbank Service Definition Language (MSDL), which is a subset of Web Service Definition

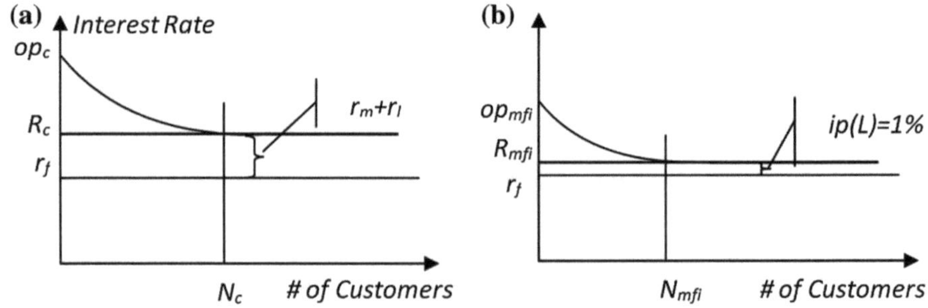

Fig. 5.4 The loan interest rate R_{mfi} of MB4ME microfinance is much lower than the loan interest rate R_c of commercial banks and converges to the minimum rate much faster: N_{mfi}. ≈ 10,000 << N_c

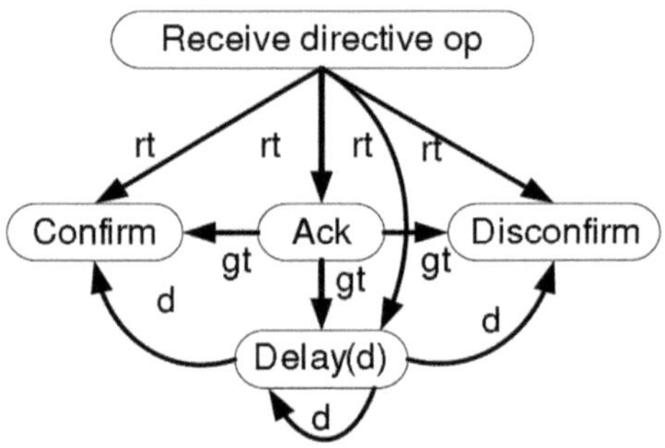

Fig. 5.5 Well behaved banking-service interface state flow diagrams for a directive one-way operation (e.g., SellItem). It is assumed that the response time, *rt*, is smaller (earlier) than the goal achievement time, *gt*: *rt* << *gt*. For SMS binding, *rt* is about 4 s

Language (WSDL), to allow interoperability between various systems over heterogeneous mobile devices, cell-phone networks, and Internet services. MSDL includes Short Message Service (SMS) binding to allow services to be discovered and utilized over SMS. To ensure that the services are usable by 2.2 billion people, the core-banking system implements a Well-Behaved Service Interface (WBSI) based on human emotion models (Song and Governatori 2006; Song and Vong 2013).

The Well-Behaved Banking Service Interfaces conforms to the above descriptions well-behaved-interfaces. Figure 5.5 shows the state flow diagram for the well-behaved-interfaces for a directive one-way operation, *op*. In the figure, upon receiving a directive operation, the service must respond within a certain response time rt_g with one of the following messages:

1. A confirming message, confirm_g, or
2. A disconfirming message, disconfirm_g, or
3. A delay message, $\text{delay}_g(d)$, or
4. An acknowledgement message, ack_g.

If the response is either *confirm*$_g$ or *disconfirm*$_g$, the communication ends. But, if the response is $delay_g(d)$ or ack_g, the user is expecting one of the above messages (except ack_g) again within a certain goal achievement time, gt_g. If the following response is $delay_g(d)$, this time the user is expecting one of the above messages (except ack_g) within the delay time d.

In order to develop a banking system for microfinance, the following main design goals were set focusing on coverage, cost, and usability:

- The system should be usable by any region with 2G cell phone networks, as 3G services are currently limited to metropolitan areas. This will ensure that we can cover 79 % of the total developing world population.
- The banking system must be affordable for most developing countries, such as Bangladesh.
- The services should be usable (effective, efficient, easy to learn, helpful and convenient) for any mobile phone users.

The whole microfinance system must be easily operable by anyone, including mobile branches (e.g., mobile operators on motorbikes. The villagers, merchants (or sellers), and microfinance bank officers should be able to perform the banking services (e.g., account creation, balance checking, lending, and money transfers) using any 2G mobiles phone capable of Short Message Service (SMS). For this, we have decided to implement the whole core-banking server on a smart phone, which can automatically process SMS messages to provide banking services over SMS. The server then can be powered by batteries and carried on motorbikes by mobile bankers. The villagers, merchants, and microfinance bank officers can use the banking service by using any 2G mobiles phones capable of Short Messaging Service (SMS).

Let us assume that the mobile banking system comprises two types of devices: mobile bank servers operating on mobile devices capable of running a program for processing electronic messages as well as sending and receiving electronic messages (e.g., SMS) and mobile stations (e.g., customers and mobile operators) that can send and receive SMS messages. With this assumption, we describe a banking system having only one mobile bank server using SMS for communication. Bankers, buyers, or sellers use the mobile stations. In this setting, we also assume that bankers operate the mobile bank server to manage accounts of all participants. This simple setting can provide the following basic trading services:

- Bankers can create accounts for buyers and sellers.
- Bankers can provide bank loans to buyers (lending service).
- Sellers can sell goods to buyers and receive credits on their accounts.
- Buyers can buy goods from sellers and credits can be deducted from their accounts.
- Buyers and sellers can enquire about their credit balances.

To ensure the security of the banking system, it uses one-time passcode for verification of buyer and sender identities. The face-to-face transaction takes place as follows:

1. The seller requests the phone number of the buyer and sends a credit request message to the mobile banking server.
2. The bank server sends a one-time passcode to the buyer.
3. The seller sends the one-time passcode to the bank to verify the passcode before handing over the items.

We should however note that the typical transaction amount of micro-entrepreneurs is very small, usually in few dollars or less. Therefore, the additional verification step (3) could hinder the daily business for busy markets. As credit fraud risk can be diversified when there are many customers and the risks can be absorbed by insurance company, an alternative was proposed considering this:

1. The seller requests for the phone number of the buyer and sends a credit request message to the mobile banking server.
2. The bank server then sends a one-time passcode to both the seller and buyer.
3. The seller visually verifies the passcode of the buyer with his passcode before handing over the items.
4. To cancel the transaction, the buyer can send a refute message to the bank server. The bank server then holds the transaction amount is in an escrow account, which can be resolved via a mobile teller or the seller's agreement.

The step (4) discourages both seller and buyers from doing any fraudulent transactions as no one will benefit from inappropriate behaviours. To evaluate the security risk of the proposed transaction protocol, a review of credit card transaction protocols was conducted. We found that the proposed protocol is similar to typical credit card transactions that only merchants send messages to banks for authorization. The security and trust measures of credit card (e.g., the 24 h grace period in the event of stolen or lost cards) can be provided to the insurance company for our mobile banking system to provide the same or better level of service quality, security, and trust.

The credit card transaction protocol was implemented using a stateless rule-based decision tree. Figure 5.6 illustrates a transaction. Figure 5.7 shows an overview of the mobile phone-based trading system that was used in our trial. The flow describes how account creation, lending, and buying/selling banking activities are handled using a

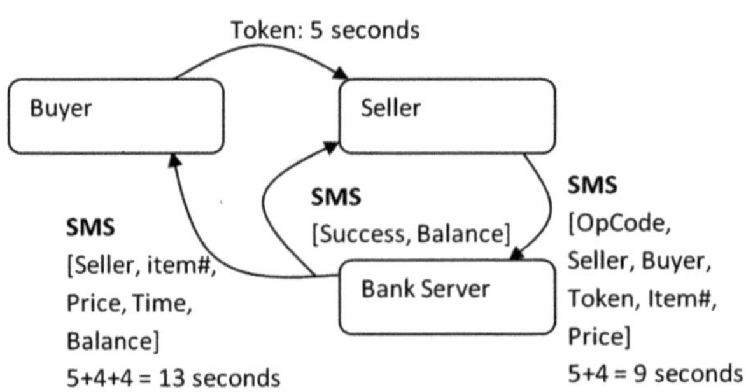

Fig. 5.6 A buyer-seller transaction. The whole transaction completes within 13 s

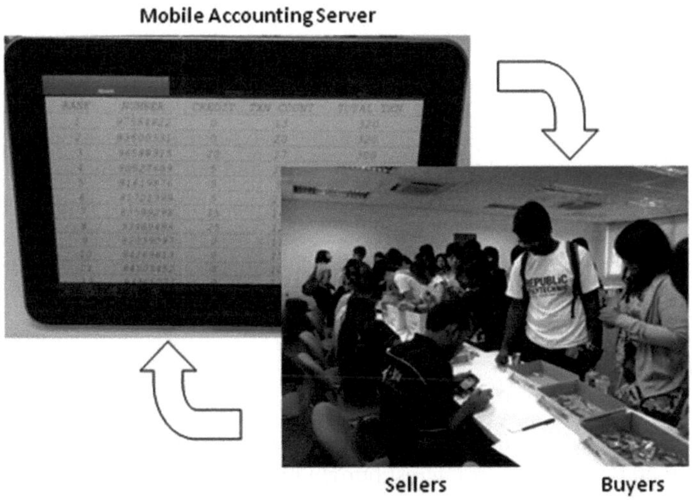

Fig. 5.7 The figure shows a seller sending an SMS message to the mobile bank server for a transaction. The server then sends authorization codes to the two parties (the buyer and seller) to complete the transaction

stateless algorithm. The server operates in a stateless mode except for the account balances: an incoming message is processed only based on account numbers, roles, and available credits. Therefore, the system is highly reliable. When the server receives an SMS message, it checks if the sender exists in the account table using the sender's mobile phone number and the sender's role. Based on the role, the server processes the message content to create an account or to credit or deduct a certain amount to/from the buyer or seller's account. Confirmation messages are also sent to the participants of the transactions. The participants will decide to trade physical goods according to the instructions sent by the bank server. Therefore, the trading system depends on the reliability of SMS messages. If SMS messages fail to be delivered within the response time, *rt*, the server must retry and inform system administrator to contact the participants. However, this would happen very rarely; therefore, its impact on the usability would be minimal.

The mobile bank server was tested on various Android phones (SAMSUNG Galaxy Note, SAMSUNG Galaxy Tab 8.9, and HTC Desire). Google Android SDK was used on Eclipse IDE using JAVA programming language with the target Android OS version 2.2. The users were able to receive authentication responses from the mobile bank server within 10 s. This delay would be the factor in influencing the service usability and it is dependable on the cell-phone infrastructure. This study strongly suggests that it is feasible to provide an affordable banking system for microfinance by utilizing only 2G cell phone networks. This is significant since the current availability of 3G networks is limited in many of the developing countries. The total cost of ownership of the banking system could go as low as US$100. The system would be able to support about 2,000 transactions (assuming 10 s per transaction limited by the current SMS services).

According to this study, the average users would feel comfortable using mobile phone-based transactions, thereby lessening doubts about whether villagers would use such a system. The mobile banking server was able to support the credit-card-like transaction protocols ensuring that users would experience the same level of security, trust, and service quality of credit card services. In comparison with a mobile money service called WING Money launched in Cambodia in 2008, (WING 2009), the protocol is much simpler, eliminating the need for a money agent. The main reason is because the mobile banking system can support not only lending, but also merchant transactions, thereby creating a cashless transactions environment.

5.6 Conclusion

In this study, we showed that among all costs incurred by MFIs, operating cost is the most important determinant of interest burden in Cambodia, Vietnam and Indonesia. We believe the same also apply to most MFIs globally. In order to lower the interest burden of MFIs, we believe that a new banking model that integrates mobile technologies and financial market innovations is required. We illustrate a possible new banking model, Micro-banking model, which uses mobile technologies to considerably lower operating cost of MFIs and increase the reach of MFIs to the poor with lower capital requirements. The integration of a micro-insurance company to the Micro-banking model will allow the pooling and distributing of risks to prevent MFIs from being overly risk averse in their selection of borrows. Therefore, the proposed Micro-banking model can help to lower interest burden, and hopefully, improve the depth and breadth of outreach of MFIs, and providing a higher access of microcredit for the poor with an interest rate close to the risk free interest rate.

Acknowledgments We would like to thank Rajendhar Dayalan, Ramesh Ramachandran, and Kevin Raj, who were involved in the evaluation and development of the mobile banking server and the JCU Singapore student volunteers who helped during the trial. The authors also would like to thank the Republic Polytechnic (RP) students who participated in the trial and the RP staff who organized RP students to participate in the trial and who conducted the survey, in particular, Leslie Sim, Elaine Li, Shamla Ramasamy, Joshua Yeo, and Sim Choon Hou.

References

Ardic OP, Heimann M, Mylenko N (2011) Access to financial services and the financial inclusion agenda—a cross-country analysis with a new data set. Policy research working paper #3357

Beshouri C, Chaia A, Cober B, Gravrak J (2010) Banking on mobile to deliver financial services to the poor. In: Mertz R (ed) Global financial inclusion. McKinsey & Company, USA, pp 24–31

[CGAP] CGtatP (2008) Foreign capital investment in microfinance: balancing social and financial returns. Focus Note 44

Chaia A, Dalal A, Goland T, Mayasudhakar MJ, Morduch J, Schiff R (2010) Half the world is unbanked. Financial access initiative. http://financialaccess.org/sites/default/files/publications/B25_Half_the_World_is_Unbanked%282011-01-21%29-1.pdf. Accessed 17 May 2010

Consultative Group To Assist The Poor/The World Bank Group (CGAP) (2010) Financial access 2010—the state of financial inclusion through the crisis. The world bank group. http://www.cgap.org/p/site/c/template.rc/1.9.47743/. Accessed 8 Dec 2011

Donner JaT CA (2008) Mobile banking and economic development: linking adoption, impact, and use. Asian J Commun 18(4):318–332

Edgar DC (2009) Realizing the full potential of mobile commerce orchestrating mobile payments and money transfers. Edgar, Dunn & Company. http://www.sybase.com/mCommercewhitepaper, 2012

Fraser DR, Gup BE, Kolari JW (1995) Commercial banking. West Publishing Company, San Francisco

Garikipati S (2008) The impact of lending to women on household vulnerability and women's empowerment: evidence from India. World Dev 36(12):2620–2642

Hammond AL, Kramer WJ, Katz RS, Tran JT, Walker C (2007) The next 4 billion: market size and business strategy at the base of the pyramid. World Resources Institute, International Finance Corporation, Washington, DC

Hughes N, Lonie S (2007) M-PESA: mobile money for the "unbanked" turning cellphones into 24-hour tellers in Kenya. Innovations 2(1–2):63–81

International Telecommunications Union (ITU) (2011) ITU world telecommunication/ICT indicators database

Kauffman RJ, Riggins FJ (2012) Information and communication technology and the sustainability of microfinance. Electron Commer Res Appl 11(5):450–468

Kiiru WK, Pederson GD (1996) Kenya women finance trust: case study of a micro-finance scheme, vol 1. Africa Region, The World Bank

Kim JB, Kang S (2012) A study on the factors affecting the intention to use smartphone banking: the differences between the transactions of account check and account transfer. Int J Multimedia Ubiquitous Eng 7(3):87–96

Kumar K, McKay C, Rotman S (2010) Microfinance and mobile banking: the story so far. Focus note 62

Pous M, Serra-Vallmitjana C, Giménez R, Torrent-Moreno M, Boix D (2012) Enhancing accessibility: mobile to ATM case study, Consumer Communications and Networking Conference (CCNC), 2012 IEEE, LasVegas, NV, pp 404–408. doi:10.1109/CCNC.2012.6181024

Rosenberg R, Gonzalez A, Narain S (2010) The new moneylenders: are the poor being exploited by high microcredit interest rates? vol 92. Emerald Group Publishing Limited, England

Song I, Governatori G (2006) Affective web service design. In: PRICAI 2006: trends in artificial intelligence, pp 71–80

Song I, Vong J (2013) Affective core-banking services for microfinance. In: Computer and information science. Springer, Berlin, pp 91–102

Tan M, Teo TS (2000) Factors influencing the adoption of internet banking. J Assoc Inform Syst 1 (5):1–42

(WING) WMC (2009) Social impact report. WING money Cambodia. http://www.wingmoney.com/fileadmin/templates/main/images/Social_Impact_Report_2009.pdf. 2010

World Bank (1999) The world bank and microenterprise finance: from concept to practice. Operations Evaluation Department, World Bank

Yunus M (1999) Banker to the poor: micro-lending and the battle against world poverty. Public Affairs, New York

Zhou T (2012) Understanding users' initial trust in mobile banking: an elaboration likelihood perspective. Comput Hum Behav 28(4):1518–1525

Chapter 6
Mobile Core Banking Server

Abstract Banking is essential to modern economies, yet nearly half of the world's people are unbanked and underserved by banks. This is mainly due to the high operating costs of banks that are passed on to retail customers, driving away customers who are not able to afford the high service costs. Without having access to the financial system the unbanked and rural poor have little means to engage in mainstream economic activities. In this chapter, we present a low-cost mobile phone-based core-banking system. This system was designed to provide retail banking services to the world's unbanked population and rural poor of 2.2 billion based on a cashless banking system that reduces operating cost and lending risks and therefore lower the current exorbitant rates charged on retail lending and micro-credit. Our analysis and evaluation shows that the mobile core-banking system on a low-cost mobile device (an Android tablet) is able to serve over 15,000 micro-banking customers. The system is therefore suitable for retail banks or microfinance institutions.

Keywords Mobile banking · Rural finance · Retail banking · Cashless payments · SMS banking · Financial inclusion

6.1 Introduction

A core banking system is a banking transaction processing system that has been extensively used by retail banks to improve efficiency and reliability of daily transactions, allowing banks to provide more products and services than before. Core banking systems also allow banks to link multiple branches using telecommunication networks to serve more customers than before. When cash is withdrawn from a bank account at an ATM machine or an item is purchased at a local grocery store, the core banking system validates the customer identity, authorizes access to the account, processes the transactions, and records the transaction details systematically to ensure that every process in the transaction is taken into account.

J. Vong and I. Song, *Emerging Technologies for Emerging Markets*,
Topics in Intelligent Engineering and Informatics 11,
DOI 10.1007/978-981-287-347-7_6

Table 6.1 Cost of core banking system

Cost items	Cost US dollars
Core system license fee	$1,200,000
Customization	$700,000
Other software license fees (please give details)	$800,000
Implementation cost	$1,000,000
Maintenance/yearly	22 % of license fee

Source IBS Intelligence (2012)

Consequently, core banking has now become synonymous with modern retail banking in the financial industry. That is, core banking system is essential to a modern economy. However, not everyone could afford products and services offered by banks. In fact, 56 % of the world's total population is still unbanked (Chaia et al. 2010). The main reasons for this phenomenon are the high cost of core banking systems and operating costs (e.g., credit underwriting and transaction). Table 6.1 shows that an entry level core-banking system can easily cost over a million US dollars. In addition, banks must also pay customization and implementation fees that often dwarf the system license cost. The complexity of these core banking systems also demands daily maintenance by highly skilled IT professionals.

The high cost is invariably passed on to customers through transaction fees and high lending rates. A third of the unbanked population, approximately 800 million people, lives in Asia. These people, who are in the lowest income category (i.e., living on under $5/day), simply cannot afford the high bank charges. It is only natural to ask why banks currently cannot provide banking services to everyone. The fundamental problem is that the current banking systems, policies, and operations are designed for the top of economic pyramid that is the wealthy mass market and the high net-worth individuals. For instance, small firms are significantly disadvantaged in financing business investment (e.g., due to 'lending bias') in many developing countries (Berger and Udell 1995; Chow and Fung 2000). Therefore, a new approach is required.

In the last decades, various solutions have been suggested, such as solidarity lending and village banking, pioneered by Grameen Bank (Yunus 1999) in Bangladesh that delivered basic banking services for the poor. Their new banking model lowered the operating costs but increased the risk of lending, and thereby increasing lending rates. These approaches attempt to bridge the gap between the two incompatible partners: the unbanked population usually the rural poor and banks that are designed to serve the wealthy. However, most micro finance institutions (MFIs) operate based on manual processes that involve posting transactions from one accounting ledger to another with human hands. The processes are not easily scalable and are susceptible to fraud and other accounting irregularities.

Our solution, called MicroBanq, is developed to fully utilize the extensive penetration of cell phone networks and 2G devices. Currently, the global mobile phone subscriptions have reached over 87 % of the global population and 79 % of developing world's population (International telecommunications union (ITU) 2011). It now covers the world's most unbanked regions. This suggests that mobile phone-based banking systems could be the ideal solution for providing banking services to those regions. MicroBanq system is mobile phone-based, cashless, branchless and wireless retail banking system. It differs from the existing other core banking systems that provide mobile phone-based banking as an additional service (whether it is "mobile phone banking" or "mobile money services" (Beshouri et al. 2010; Donner 2008)), such as Sybase mCommerce 365 by SAP (Edgar 2009), and DELL (2012). As these existing systems are targeted at developed banking sectors, their core banking systems are not only expensive for banks in emerging economies but would also incur high associated transaction cost.

MicroBanq is a new mobile phone-based core banking system. The system was developed based on a new banking model shown in Fig. 6.1. In the following sections, we describe the details of system architecture and evaluation of a prototype including more detailed motivations and backgrounds of rural banking. A prototype cashless payment of the MicroBanq system was installed on an Android tablet and trialed in a "small and artificial village" comprising five mobile bank operators, five merchants (sellers), seven companies and 130 buyers. The participants performed

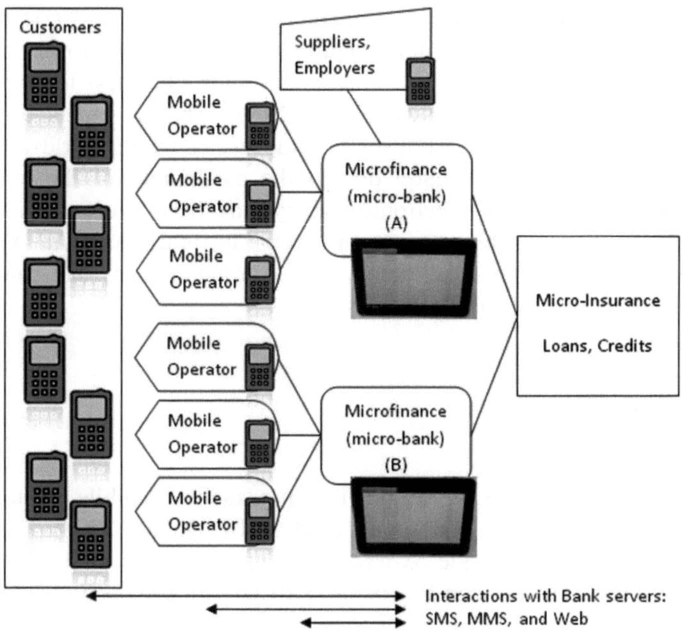

Fig. 6.1 System diagram of cashless, branchless and wireless mobile phone-based banking model. The system shows a core banking system, called MicroBanq, running on Android based tablets. Bank tellers now operate on mobile offices, such as motor bikes. The insurance company pools the risks to reduce the overall credit risk through diversification, providing low lending interest rates

804 transactions and sent 2,412 short message service (SMS) messages over a three hour period. The results show that the payment system on a low-cost Android device is able to serve over 15,000 micro-entrepreneurs, making it large enough for most micro-banking and microfinance institutions. Section 6.6 reports on these performance results, as well as usability and user experience survey results.

6.2 Background and Motivation

6.2.1 Rural Finance

World Bank (1999) defines microenterprise as "informal sector business with five workers or less, and fixed assets valued at less than US$10,000". This definition is similar to that used by USAID (Blayney and Otero 1985) and others (Ayyagari et al. 2007). Provision of financial services in rural area always poses challenge for banks operating in the developing economy {World Bank, #129}. In developing countries, the majority of micro-entrepreneurs is self-employed or hires employees without salaries, while hoping for paid employment (Tokman 2007). For instance, a third of the non-agricultural labour sector is operated by the self-employed. Similarly, a half of the informal sector is operated by self-employed (De Mel et al. 2010). The main three reasons for micro-entrepreneurs being in informal sectors are lack of access to credit, inability to obtain approval for a business license, or being prevented from starting a business (De Soto 1989). The movement by Grameen Bank (Yunus 1999) in Bangladesh inspired rural finance efforts world-wide to help these micro-entrepreneurs. Recently due to the availability of the internet and telecommunication services, electronic money has been proposed as possible solutions for empowering the poor. A study done in rural Cambodia (Vong et al. 2012) showed that the financial access to electronic money has changed and improved livelihoods of rural Cambodia.

6.2.2 Mobile Banking

Various terms have been used for mobile banking, such as "mobile phone banking" and "mobile money services" (Beshouri et al. 2010; Donner 2008). Examples of mobile banking systems include Sybase mCommerce 365 by SAP (Edgar 2009) and DELL Mobile Banking and Payments (DELL). In developed countries, much research has been conducted recently into development of mobile phone-based banking accessible to people with disabilities (Pous et al. 2012) and security and trust issues (Zhou 2012) as well as public perceptions of and attitude toward such systems (Kim and Kang 2012).

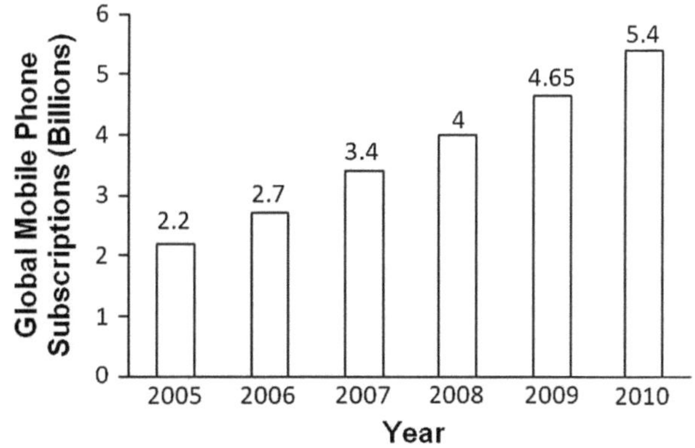

Fig. 6.2 Number Of global mobile phone subscriptions. *Source* ITU World Telecommunication/ICT Indicators Database (2011) (http://www.itu.int/ITU-D/ict/statistics/index.html)

6.2.3 Unbanked Population

A report commissioned by World Bank Group/World Resources Institute estimated that about 80 % of the world's population (4 billion people) still have annual incomes below the equivalent of US$3,000 in local purchasing power (Hammond et al. 2007). The ITU reported that global mobile phone subscriptions accelerated from 2.2 billion in 2005 to over 5 billion in 2010, reaching 87 % of the total global population and 79 % of the total developing world population (see Fig. 6.2). This penetration was set to grow to 6 billion by end 2011 (International Telecommunications Union (ITU) 2011). Chaia et al. (2010), CGAP (2010) and Ardic et al. (2011) suggested that by end 2009, there would be an estimated 2.75 billion people out of 5 billion people (that is, 56 % of global adult population) without access to formal financial services for saving, borrowing, or transacting.

Very recently, the mobile phone has been widely used as a channel for providing retail banking, microbanking and microfinance services, hence the terms "mobile phone banking" and "mobile money services". Furthermore, the use of mobile phones to provide financial services across developing countries has been one of the most remarkable technology stories in the past decade. By end 2009, more than 120 mobile money projects were deployed in 70 emerging countries (Beshouri et al. 2010). This phenomenon has since been given impetus by a range of terminologies such as mobile banking (m-banking), mobile payments (m-payments), mobile transfers (m-transfers), and mobile finance (m-finance) (Donner 2008).

6.3 MicroBanq System Architecture

In this section, we describe MicroBanq system architecture in detail. MicroBanq is a mobile phone-based core banking system, specifically designed to lower the cost of ownership for MFIs. The latest MicroBanq system currently costs under US $500, which includes an Android based tablet specifically configured for the core banking system. MicroBanq system also includes MicroGuard, which is a set of information security policies, standards, practices and procedures to ensure availability, consistency, and integrity of the core banking system. The information security standard is based on the recommendations of National Institute of Standards and Technology (NIST) and Information Security Forum (ISF).

MicroBanq is available from IdeasBanq Singapore or partners. IdeasBanq also provides customization, implementation, and maintenance services at low cost. The overall system architecture is shown in Fig. 6.3. As you can see from the figure, MicroBanq uses two-factor authentication used by other modern core-banking systems: PIN and SMS. This architecture is based on the mobile core banking system model shown in Fig. 6.1. MicroBanq system has been extensively tested and studied by various researchers in the past (Song and Vong 2013a, b; Song et al. 2014; Lui et al. 2013).

A typical MFI has one manager who understands the overall operation, policies, standards and practices to follow that meets the regulations and ethical and cultural requirements of the local community. The manager is in charge of training and educating all employees of the policies to meet the obligation of the organization: due care and due diligence. Therefore, the manager has the highest authority and governs ownership, access, and control of all information assets of the organization. A typical MFI also has one or two IT support personals for day to day maintenance of office computers. An MFI would also have a dedicated cashier or customer service officer (CSO) depending on the workload. Figure 6.3 illustrates how an MFI can utilize a mobile phone-based core banking system. The core banking system

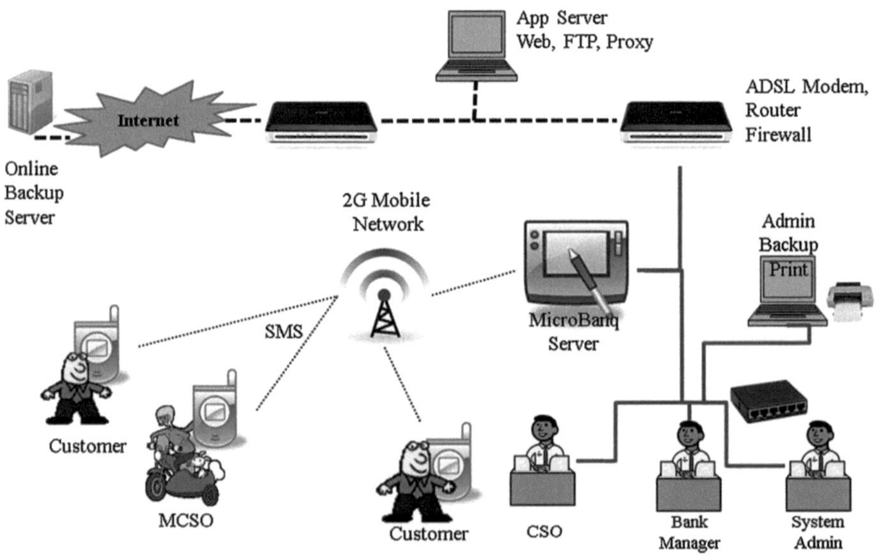

Fig. 6.3 An overview of MicroBanq system. Customers and MCSO access MicroBanq server via SMS. CSO and Bank manager access the server using a client software running on a computer or a mobile phone over HTTP protocol within the internal network of the MFI

(MicroBanq) is installed on an Android tablet. The bank manager may directly access the system by interacting with the tablet or through a personal computer (PC) via their internal network. CSOs can access the core banking server only using a PC via their internal network. A system administrator is responsible for maintaining the internal networking, internet connections, and information security incident discovery, response, and recovery according to the recommendations of MicroGuard.

Customers access their accounts conveniently using their mobile phones anywhere anytime. Mobile customer service officers (MCSOs) visit customers living in hard to reach rural areas to collect basic identity information of customers, accept cash deposits. The basic idea behind the mobile banking model (Fig. 6.1) and MicroBanq architecture (Fig. 6.3) is that, by eliminating cash, providing mobility of services to cover hard to reach areas through 2G networks, we can greatly reduce the associated risk, resulting and reduce cost. This can be passed to the customers as lower loan interest rates and lower service charges. Chapter 5 analysed the cost model in detail and further elaborated why this is possible.

Figure 6.4 illustrates the interactions between actors (CSO, MCSO, Cashier, Manager, fund line, Customers, and MicroBank server) and message types (SMS, voice, HTTP, and physical cash). The main system design revolves around the interaction diagram and the accounting components shown in Fig. 6.5. For brevity, Fig. 6.5 shows only savaging account and loan account components. However, MicroBanq can be configured to include many other financial products in modular approach. In addition, it comes with the following built-in modules: accounting (e.g., purchase orders handling and taxation), payroll, and customer management, and daily cash balance and transaction reports.

Access to information assets are tightly controlled based on MicroGuard to safeguard the assets and to ensure business continuity when security incidents are identified. Table 6.2 shows sample access control configuration for information assets of MicroBanq system.

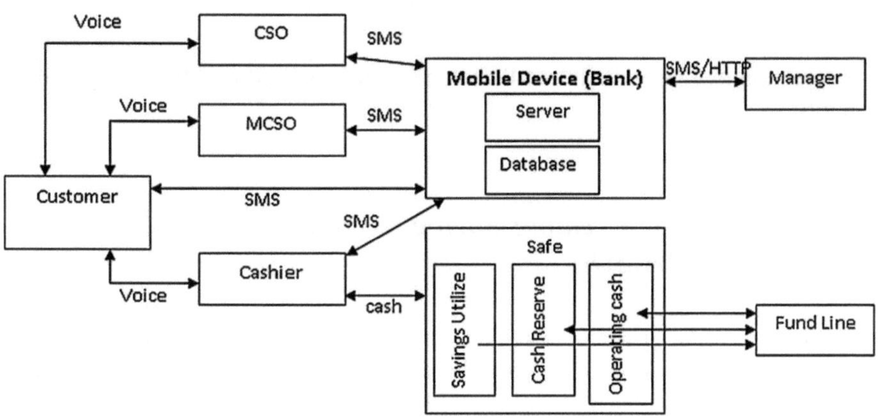

Fig. 6.4 MicroBanq system interaction diagram shows the interactions of different components of the mobile core-banking server: customer, CSO, MCSO, bank manager, cashier, cash safe, and fund line

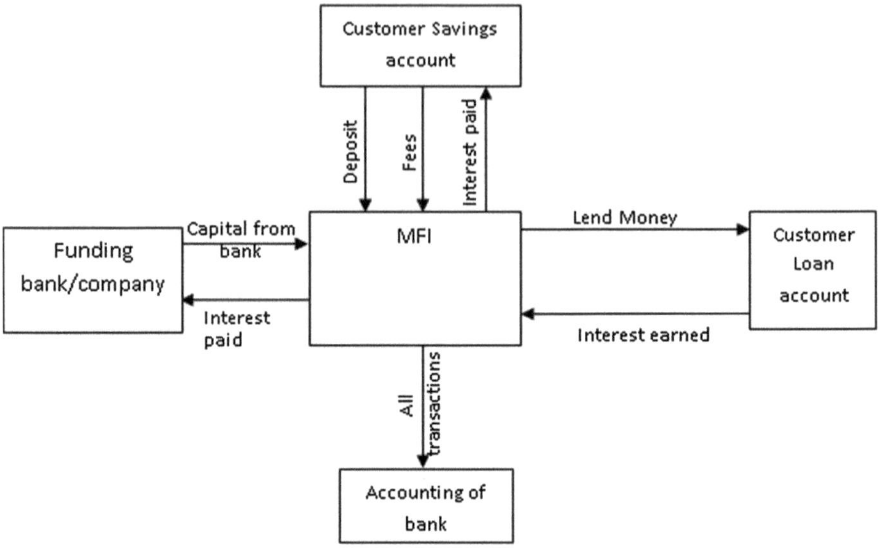

Fig. 6.5 Overview of the accounting components of the mobile MFI

To work in a secure environment the MB needs to define a role for each user. Each user is tied to a role and its permissions scheme (that defines all its accesses and privileges). In the proposed system the users are: Manager (BM), Cashier (CH), MCSO, CSO and Customer (CT). Customers are further classified into Buyer (BR), Seller (SR), and Corporations (CP).

6.4 Usage Scenarios of MicroBanq System

In this section, we illustrate how MicroBanq system is used by MFIs using various scenarios.

6.4.1 Scenario S1: Creating Saving Account Through MCSOs

A customer (e.g., a farmer, a merchant, an employer), who does not have a bank account or have a bank account somewhere else, wants to know about the financial services provided by the MicroBanq MFI. The customer calls a CSO using his mobile phone to enquire about the services and benefits of Savings Account (SA) product. After the consultation, the CSO sends the details of the customer's enquiry to the server for record keeping.

Table 6.2 Users, roles and access levels

Information assets	Description of action	Access
Employee records	Add/delete/update new employ details	BM
Access levels	Update the access levels of employees	BM
Savings account products	Updating the default values related to saving accounts	BM
Fund lines	Adding/updating a new funding source	BM
Installment payments for fund line	Paying installment to the funding source	BM
Savings accounts	Create a savings account for customer	BM, CH, MCSO
Loan accounts	Create a savings account for customer	BM, CH, MCSO
Last event of an agent	Cancel the latest event (repayment, disbursement, reschedule, all types of events)	BM
Manual interest rate	User is able to enable manual calculation of interests when repaying a loan	BM, CH, MCSO
Manual fees calculation	User is able to enable manual calculation of fees (commissions and late fees) when repaying a loan	BM, CH, MCSO
Close and withdraw	User is able to close and withdraw remaining balance from a savings account	BM, CH, MCSO
Close and transfer	User is able to close and transfer remaining balance from a savings account	BM, CH, MCSO
Transfers between accounts	User is able to perform a transfer from one savings account to another	BM, CH, MCSO, CSO
Withdraw	User is able to withdraw from a savings account	BM, CH, MCSO, CSO
Deposit	User is able to perform a deposit on savings account	BM, CH, MCSO, CSO
Buy/sell	User is able to perform buy/sell operation	BM, CH, MCSO, CSO, customer
Bills	User is able to pay a bill to other organization	BM, CH, MCSO, CSO
Installment payments	User is able to pay installment on loan account	ALL
Customer information	User is able to create customer information	BM, CH, MCSO, CSO
View customer information	User is able to view customer information	ALL
Update customer information	User is able to update customer information	BM, CH, MCSO
View G/L entry	User can view general ledger entry	BM, CH, MCSO
Enter G/L entry	User can enter new general ledger entry	BM
View report on cash flow	User can view the daily cash flow	BM, CH, MCSO
View balance sheet	User can view the daily balance sheet	BM, CH, MCSO

The customer, after enquiring about the product, requests the CSO to open an account for the SA product. The CSO then send the phone number of the customer to a MCSO via SMS.

The MCSO then arranges a schedule with the customer and meets with the customer. MCSO collects customer information (CI) for SA (e.g., name, address, national ID, certificate from the head of the village). MCSO verifies the information and sends a basic customer information (BCI) comprising of those details and an initial deposit amount to the server. The server then notifies the customer of the account details and current balance of the account via SMS.

MCSO later returns to the MFI and deposits the initial deposit amount he has collected from the customer at CSO.

6.4.2 Scenario S2: Creating Saving Account Through Customer Service Officers at MB

In this scenario, the customer visits the MicroBanq MFI to obtain information about the SA product from a CSO and request to open an account for the product. The customer can deposit an initial deposit amount at CSO. The CSO collects the Basic Customer Information (BCI) and creates a savings account. The server notifies the customer of the details of the new account via SMS.

6.4.3 Scenario S3: Saving Account Cash Deposit Through MCSO

A customer calls CSO and state that he wishes to deposit some cash. The CSO sends an SMS message to MCSOs. An MCSO, who is near the customer, responses to the job. The MCSO then visits the customer and receives the cash from the customer. MCSO notifies the server the information (customer account and the deposit amount) of the deposit via SMS. Server informs customer the updated balance. MCSO later deposits the collected cash at a cashier. The cashier informs server. The server updates the balance of MCSO.

6.4.4 Scenario S4: Saving Account Withdrawal Through MCSO

A customer wants to withdraw some cash. The customer goes to a CSO and requests for a withdrawal of a certain amount of cash. CSO issue a withdrawal request to the server specifying the amount. The server verifies whether the requested amount is available or not. If available, the server sends a random verification code to both the customer and the CSO. The CSO then matches the verification code, gives the requested amount of cash to the customer, and send a notification to the server. The server informs the updated balance to the customer.

6.4.5 Scenario L1: Provide Loan Information

A customer wants to take a loan from the MicroBanq MFI for some purpose (e.g., to buy fertilizer).

The customer first calls the CSO to gather information about taking a loan, interest rates, and documents required. The CSO requests the server to send the loan information to the customer for loaning a certain amount for a certain number of days. The server calculates interest rates and sends information on the loan interest rates, eligibility criteria, loan options, and payment options via SMS.

6.4.6 Scenario L2: Loan Application Through MCSO

An MCSO visits a customer who wants to loan a certain amount for a certain number of days. The MCSO receives a list of documents and information that need to be collected and verified. MCSO collects the documents and information of the customer and sends them to the server. The server calculates credit rating and interest rates based on the credit ratings of the customer. The server sends this information to the MCSO.

The customer agrees to the terms and signs a loan application form. MCSO sends a request for the loan application. The server sends the new loan account information to the MCSO and the customer via SMS. The MCSO gives the loan amount to the customer.

6.4.7 Scenario L4: Instalment Payment By Cash Through MCSO

This is similar to the scenario S3 (saving account cash deposit through MCSO). An MCSO visits the customer to collect an instalment of a loan. The MCSO sends a request (for the instalment collection information for the customer) to the server. The server sends the amount to be collected to MCSO and the customer.

The MCSO collects the cash and sends a request for instalment payment to the server. The server informs the balance of the loan account to the customer and the MCSO. The MCSO deposits the collected amount at a cashier at the MicroBanq MFI. The cashier notifies the server of the deposit by the MCSO. The server updates the balance of the MCSO and notifies the cashier.

6.4.8 Scenario T1: Merchant Sell

A buyer (a customer of a MicroBanq MFI) goes to a seller (another customer of the MicroBanq MFI) to buy some items. The buyer tells his account number to the seller.

The seller sends the buyer's account number and price of the items to the server via SMS. The server checks the current balance of the customer. If the balance is greater than or equal to the requested amount, the server sends an approval message to seller and the buyer via SMS. The seller hands in the items to the buyer. The buyer collects the items.

6.5 Interaction Design of Mobile Core-Banking Server

As illustrated in Chap. 5, the core-banking system defines and exports its services through microbank service definition language (MSDL), which is a subset of web service definition language (WSDL), to allow interoperability between various systems over heterogeneous mobile devices, cell-phone networks, and internet services. MSDL includes SMS binding to allow services to be discovered and utilized over SMS. well-behaved service interface (WBSI) based on human emotion models is implemented to allow usability of the services by 2.2 billion people (Song and Vong 2013a; Song and Governatori 2006). To ensure the security of the banking system, it uses one-time passcode to verify buyer and sender identities. More detailed description of the interaction design is described in Sect. 5.5.

6.6 Evaluation of the Mobile Banking Server

The mobile banking system shown in Figs. 6.1, 6.3, and 6.4 was implemented on an Android tablet and evaluated on an artificial village comprising of five mobile bank operators, five merchants (sellers), seven companies and 130 buyers. The buyers were fresh high-school graduates and most of them have never used mobile banking services. The participants performed 804 transactions and sent 2,412 SMS messages over a three hour period as shown in Fig. 6.6. At around 11am (shown as "1090" in the figure), the buyers arrived to the village and the bankers created accounts for the buyers for the first time. From 12pm, the buyers were then participated in activities organized by the seven companies to earn credits and use the earned credits to purchase items from sellers. After the trial, the participants were asked to complete a questionnaire to evaluate perceived usefulness, easy-of-use, and safety.

Figure 6.7 shows the survey results. Except question 2, all responses were statistically significant ($p < 0.01$) indicating that the mobile banking system is

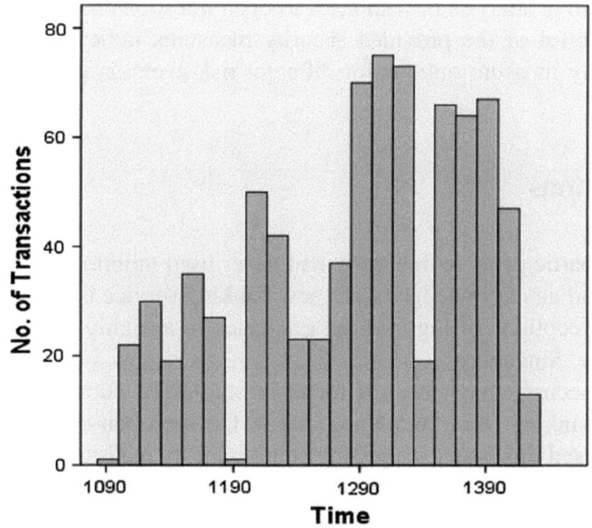

Fig. 6.6 Each bar represents the number of transactions in 10 min intervals. It shows that the server could easily handle over 75 transactions per 10 min (7.5 transactions per minute). Each transaction consists of three SMS messages

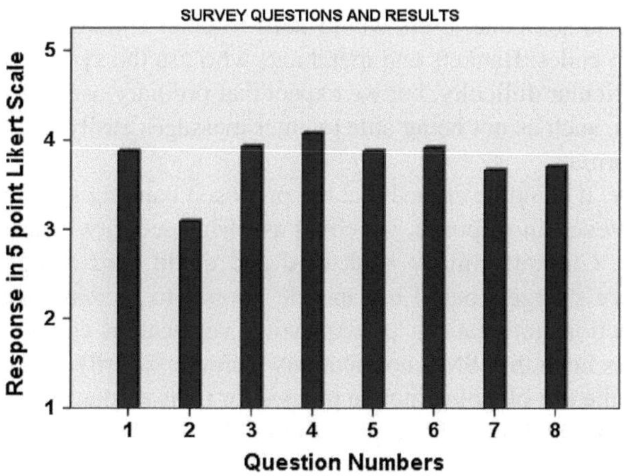

Fig. 6.7 The answers were based on 5 point Likert scale: *1* Strongly disagree, *2* Disagree, *3* Neutral, *4* Agree, *5* Strongly agree. Survey questions: *Q1* The mobile trading system is easy to use. *Q2* I think that I would need the help of other people to be able to use this mobile. *Q3* The steps required to perform mobile trading were easy to follow. *Q4* I was able learn to use this system very quickly. *Q5* Most people would learn to use this system very quickly. *Q6* I am confident that I would remember how to use the system again. *Q7* I would like to use this system again. *Q8* I would recommend this system to others

usable. Qualitative analysis of responses to open questions suggest that some users were still doubtful of the provided security measures indicating that we need to provide security measure options for different risk-aversion profiles of customers.

6.7 Limitations

Although the participants in this study had never used mobile phone-based banking services nor had any knowledge of the new banking service being tested, this study reflects the perceptions of high school graduates in a highly wired and connected urban city area, Singapore.

Trust and security have been a focus in studies of adoption rates of mobile phone-based banking (Kim and Kang 2012). On the open-ended question, a few participants raised this issue, suggesting better security measures; however, this was not a major concern for most participants. This could be due to the similarity of the transaction protocol to that of credit cards, participants' exposure to credit card system, or the perfection of credit card transaction protocol. Further study is required on this behaviour.

One critical limitation of 2G based mobile banking is the usability of 2G mobile phones. 2G mobile phones have numeric keypads laid out in a 4 × 3 grid for both dialling and messaging. Enquiring for a balance would not be a problem for normal users as they could just send any message for a balance statement, but in order to transfer money to someone, users would need to enter counterpart account details and verification codes. Bankers and merchants who use the system daily would not experience particular difficulty, but we expect that ordinary users would experience usability issues, such as not being able to enter messages easily and having to learn the message formats.

Additionally, it could be argued that the proposed banking system lacks security measures. However, in response, we could ask what security measure credit cards currently have. Currently, many bank card and credit card companies including DBS (Singapore's largest bank) use mobile phones to increase security (e.g., by sending transaction notifications or requesting verification codes). These institutions seem to believe that SMS notifications increase security. That said, methods for preventing the use of stolen mobile phones for transactions present an important future topic in research on mobile phone-based banking or mobile phone assisted banking.

Acknowledgments We would like to thank Rajendhar Dayalan, Ramesh Ramachandran, and Kevin Raj, who were involved in the evaluation and development of the mobile banking server and the JCU Singapore student volunteers who helped during the trial. The authors also would like to thank the Republic Polytechnic (RP) students who participated in the trial and the RP staff who organized RP students to participate in the trial and who conducted the survey, in particular, Leslie Sim, Elaine Li, Shamla Ramasamy, Joshua Yeo, and Sim Choon Hou.

References

Ardic OP, Heiman M, Mylenko N (2011) Access to financial services and the financial inclusion agenda–a cross-country analysis with a new data set. Policy Res Working Pap 3357:21

Ayyagari M, Beck T, Demirguc-Kunt A (2007) Small and medium enterprises across the globe. Small Bus Econ 29(4):415–434

Berger AN, Udell GF (1995) Relationship lending and lines of credit in small firm finance. J Bus 68(3):351–381. doi:10.1086/296668

Beshouri C, Chaia A, Cober B, Gravrak J (2010) Banking on mobile to deliver financial services to the poor. In: Mertz R (ed) Global financial inclusion, McKinsey & Company, USA, pp 24–31

Blayney RG, Otero M (1985) Small and micro-enterprises: contributions to development and future directions for AID's support. Agency for International Development, USA

Chow CKW, Fung MKY (2000) Small businesses and liquidity constraints in financing business investment: evidence from Shanghai's manufacturing sector. J Bus Ventur 15(4):363–383. doi:10.1016/s0883-9026(98)00014-7

Chaia A, Dalal A, Goland T, Jose M, Mayasudhakar, Morduch J, Schiff R (2010) Half the world is unbanked. Financial access initiative. http://financialaccess.org/sites/default/files/publications/B25_Half_the_World_is_Unbanked%282011-01-21%29-1.pdf. Accessed 17 May 2010

Consultative Group to Assist the Poor/The World Bank Group (CGAP) (2010) Financial access 2010–the state of financial inclusion through the crisis. The World Bank Group. http://www.cgap.org/p/site/c/template.rc/1.9.47743/. Accessed 8 December 2011

De Mel S, McKenzie D, Woodruff C (2010) Who are the microenterprise owners? Evidence from Sri Lanka on Tokman versus De Soto. International differences in entrepreneurship. University of Chicago Press, Chicago, pp 63–87

De Soto H (1989) The other path: the invisible revolution in the third world. Harper and Row, New York

DELL (2012) Mobile banking and payments. DELL. http://i.dell.com/sites/doccontent/business/solutions/engineering-docs/en/Documents/mobile-banking-payments.pdf, 2012

Donner JaT CA (2008) Mobile banking and economic development: linking adoption, impact and use. Asian J Commun 18(4):318–332

Edgar DC (2009) Realizing the full potential of mobile commerce orchestrating mobile payments and money transfers, Edgar, Dunn & Company, Singapore. http://www.sybase.com/mCommercewhitepaper, 2012

Hammond AL, Kramer WJ, Katz RS, Tran JT, Walker C (2007) The next 4 billion: market size and business strategy at the base of the pyramid. International Finance Corporation, World Resources Institute, USA

IBS Intelligence (2012) Core banking systems cost benchmark. IBS Intelligence. https://www.ibsintelligence.com/about-the-core-banking-systems-cost-benchmark-2012. Accessed 9 August 2014

International Telecommunications Union (ITU) (2011) ITU world telecommunication/ICT indicators database

Kim JB, Kang S (2012) A study on the factors affecting the intention to use smartphone banking: the differences between the transactions of account check and account transfer. Int J Multimedia Ubiq Eng 7(3):87–96

Lui C, Song I, Vong J (2013) Lower the interest burden for microfinance. In: Proceedings of the international conference on managing the asian century, Springer, Singapore, pp 185–191

Pous M, Serra-Vallmitjana C, Giménez R, Torrent-Moreno M, Boix D (2012) Enhancing accessibility: mobile to ATM case study, pp 404–408

Song I, Governatori G (2006) Affective web service design. PRICAI 2006 trends in artificial intelligence. Springer, Berlin, pp 71–80

Song I, Lui C, Vong J (2014) Lowering the interest burden for microfinance. Int J Process Manag Benchmarking 4(2):213–229

Song I, Vong J (2013a) Affective core-banking services for microfinance. In: Computer and information science. Springer, Berlin, pp 91–102

Song I, Vong J (2013b) Mobile core-banking server: cashless, branchless and wireless retail banking for the mass market. In: 2013 international conference on IT convergence and security (ICITCS), IEEE, pp 1–4

Tokman VE (2007) Modernizing the informal sector. UN/DESA working paper, vol 42, pp 1–13

Vong J, Fang J, Song I (2012) Delivering financial services through mobile phone technology: a pilot study on impact of mobile money service on micro–entrepreneurs in rural Cambodia. Int J Inf Syst Change Manag 6(2):177–186

World Bank (1999) The world bank and microenterprise finance: from concept to practice. Operations Evaluation Department, World Bank, USA

Yunus M (1999) Banker to the poor: micro-lending and the battle against world poverty. PublicAffairs, New York

Zhou T (2012) Understanding users' initial trust in mobile banking: an elaboration likelihood perspective. Comput Human Behav 28(4):1518–1525

Part II
Health

Chapter 7
Automated Health Care Services

Abstract In this chapter we would like to share new emerging technologies that are changing the way we access medical care. Despite the rural development and modernization of the way of life, the cost burden of medical care and access to medical care has become a central concern for both developed and underdeveloped countries. The number of people admitted to hospitals and medical expenditures have doubled over the past 10 years. Since the invention of radio, television, and computer, various attempts such as telemedicine have been made for improving access to medical care. Telemedicine is still being actively developed and utilized in developed countries, but its main purpose is to allow medical specialists to remotely diagnose or operate on patients. However, our focus here is the other aspect of accessibility: reducing the cost of medical care by using computerized medical services. We can broadly categorize computerized health care services into four categories: (1) tele-health, (2) automated diagnosis and assessment of health, (3) online health support, and (4) health information management systems. We review the latter three approaches: automated health care systems, highlighting the potential power of each in solving current health care problems.

Keywords Health informatics · Online health · Health social network · eHealth · Medical data analysis · Automated diagnosis

7.1 Introduction

We are now witnessing a paradigm shift in medical care. The conventional doctor-patient approach to delivering medical care is no longer sustainable. In the near future, we will no longer need to travel or visit doctors for medical diagnosis, to receive medications or treatments. Furthermore, human doctors will gradually be replaced by smart medical systems that will provide diagnosis, drugs and treatments for 24/7 extended hours, helping all people equally. We call this inclusive medical

J. Vong and I. Song, *Emerging Technologies for Emerging Markets*,
Topics in Intelligent Engineering and Informatics 11,
DOI 10.1007/978-981-287-347-7_7

care (IMC). We will see in this chapter some examples of smart medical systems that have already been deployed. Inclusive medical care is also the only way towards sustainable medical care if we believe all lives have equal value. Currently, the cost to train a general practitioner is over US$300,000. To train a specialist, it would cost over US$800,000 to society (Rothkrantz 2011). However, one practitioner can care for only a handful of patients a day. Consequently, only a select few can afford to have proper medical care under the conventional doctor-patient approach. We would like to highlight that IMC is a completely different approach from the so-called *medical home*, also referred to as patient oriented care (Collaborative P-CPC 2007) that have been proposed in the developed world. Medical home focuses on improving the quality of medical care for those who can afford to pay the increased costs rather than reducing costs for those who cannot afford the costs.

The main catalyst of this paradigm shift is the rapid global penetration of mobile network. Mobile network provides two-way point-to-point communications. More than 90 % of the global population now owns at least one cell phone whereas fixed-telephone subscription rate is still below 17 % (ITU World Telecommunication 2014) as shown in Fig. 7.1. In 5 years, most Africans will have cell phones (Badlani 2012). In addition, cell phones now have capability to capture high quality data for medical diagnosis.

Big data is the next important emerging technology that can improve health care. All those data that are captured need to be analysed in time and at low cost. Conventional approaches of storing and managing data is not scalable to hold the massive amount data generated. The world now produces more than 5 EB every 2 days. Kaiser, a hospital in the United States, alone has about 30 PB of data. The amount of data produced doubles every 2 years and the pace is accelerating. Kaiser has begun analysing some of their data only recently. Conventional relational database management systems and grid computing technologies are not suitable for this magnitude of data. It is not just because of the amount of the data, but also because we need to be able to process many different types of data. Big data processing technologies such as the Apache Hadoop are designed to store and process heterogeneous data across clusters of computers. The key to the success of this approach is scalability and reduced costs for processing large amount of data.

The aim of this chapter is to define what inclusive medical care is, classify current practices that meet the criteria of inclusive medical care, and illustrate dozens of technologies and pilot projects that can help realizing inclusive medical care.

7.2 Inclusive Medical Care (IMC)

The objective of inclusive medical care is to provide quality medical care at affordable costs. Therefore, we will first need to define the standards for quality medical care that is accessible to everyone. Secondly, we will need to define the standards for affordable costs. The medical home community has been trying to

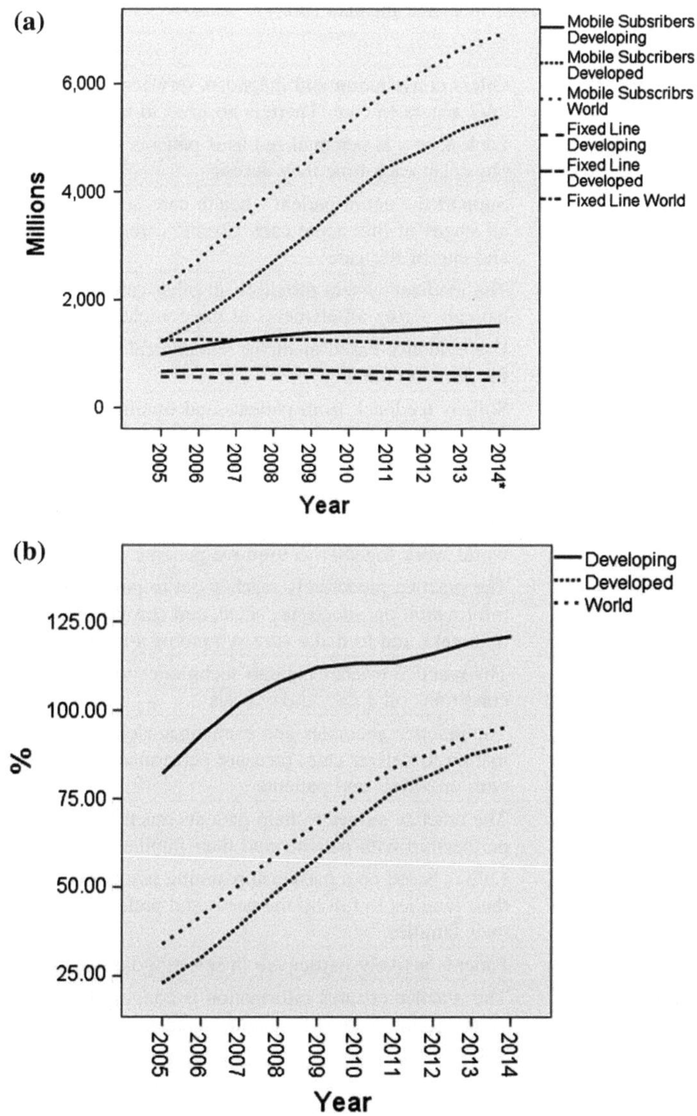

Fig. 7.1 Mobile phone subscription rate in the world

define what are considered as medical home, a patient-centred accessible medical care. As the standards developed for medical home is to improve accessibility of medical care, we can borrow some of the criteria that they have developed over the last three decades (Collaborative P-CPC 2007; AAFP et al. 2011) and refine such that the care is affordable for everyone. Table 7.1 lists the refined categories of activities for IMC. The standard highlights that the service should be available

Table 7.1 Standard activities of inclusive medical care

Activities	
Continuous access	Offers consultation and diagnosis services over extended hours: 24/7 access to care. There is no need to make appointments
Personalized care	Each access is personalized as if patients see the same personal physician each time they access
Complete care	Support the entire patient's health care needs including care for all stages of life; acute care, chronic care, preventive services, and end of life care
Integrated care	The medical system monitors all other care received by their patients across all elements of the complex health care system
Best practices	Use evidence-based medicine and clinical decision-support tools for decision making
Patient feedback	Solicits feedback from patients and families to meet patients' expectations and to facilitate quality improvement
Performance measurement	Use of performance measurement technologies to continuously optimize patient care
Collaborative care	Medical systems, researchers, patients, and practitioners in the world work together to improve patients' care
Population management	The practice proactively reaches out to people with critical information on infectious, acute, and chronic diseases to prevent outbreaks and to make sure symptoms are under control
Patient self-management	The practice teaches patients techniques to manage their conditions on a day-to-day basis
Health IT	The practice generates and exchanges electronic health information to deliver care, measure performance, and communicate with providers and patients
Care plans	The practice strives to help patients reach goals defined in partnership with patients and their families
Patient-centered care	Care is based on a partnership among practitioners, patients, and their families to full fill the needs and preferences of patients and their families
Shared decision-making	Patients actively participate in selecting treatment options
Cultural competency	The practice ensures information is conveyed to patients in a language and method they understand, taking cultural differences into account
Quality measurement and improvement	There must be an officer who is in charge of a group of patients or a region and is held accountable for performance
Automate care	Use advanced telecommunication and smart technologies to automated diagnosis, prevention, treatment, and manage to reduce cost
New payment systems	The practice offers new options for payment (e.g., cashless mobile payment)

Refined from Collaborative P-CPC (2007), AAFP et al. (2011)

continuously throughout the day and night. The care should also be personalized and care plans are managed for all stages of life. There should also be a clearly defined responsibility of the operation and performance of the care. Importantly, the overall care must be automated using smart technologies reducing the cost and allowing patient to pay remotely using methods such as mobile banking.

7.3 Online Medical Knowledgebase

The internet hosts much greater knowledge than all medical practitioners in the world could accumulate in their life time. The amount of information that is stored and replicated in the world in 2010 is estimated to be 1.2 million petabytes (Gantz and Reinsel 2010). A petabyte is a million gigabytes. And the pace is accelerating, doubling every 2 years. One can now use internet search tools to access the vast amount of data by simply entering relevant keywords, filtering, and narrowing down documents to find relevant documents that contain vital information on prevention, diagnosis guidelines, and treatment information.

Here is one scenario that highlights how *online medical knowledgebase* is changing the world. Suppose a patient Bob suffers from a chronic skin condition. The symptoms are shading skins and itchiness. So, Bob enters the following keywords into a web search tool such as Google: “skin problem, shading skins, itching, and natural remedy.” Then Bob will be presented with vast amount of information relevant to the skin condition. Bob can use the information to diagnose and order non-prescription medicines for treatment. But, Bob soon realizes that there is simply too much information. Which treatments are the best, and which will actually work for Bob?

Health Social network addresses these problems. Health social network is a place where people share their experiences on various medical conditions, treatments that worked for them, how they managed their emotional and psychological problems, where they got supports, and how they managed their care. The internet contains too much information that is usually created by entities with various interests of their own. People in the world try out various products advertised by the entities and share their stories, filtering out the vast amount of information. For example, Bob with the skin problem could immediately find that a story of a grandmother who used apple cider vinegar to treat her granddaughter’s chronic skin condition with similar symptoms, how magically the commonly available apple cider vinegar cured the age long skin condition that medical practitioners had failed to help. Only if the medical practitioners had bothered to use the Internet, the granddaughter would have suffered less. But the bigger problem is that the apple cider vinegar, which can be purchased in a local supermarket at around $1 a bottle, is not covered by medical insurance, so that it is not listed in the medical practitioners’ computers.

Let us summaries what has happened here. Figure 7.2 shows that companies and various interested parties create and publish information on various treatment

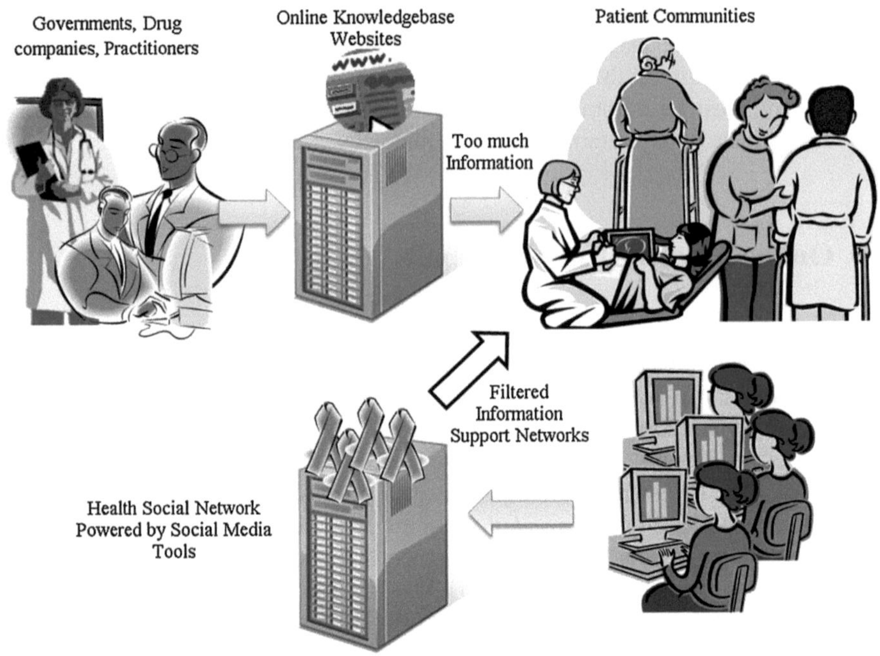

Fig. 7.2 Information filtering role of health social network

options, drugs, herbs, supplements, trainings, and clinics on the internet. However, that information is often not filtered and patients find it hard to relate with their own personal conditions. This unfiltered information is mostly about diseases, symptoms, and general treatment options. Patients however look at their conditions in the context of their daily lives and activities. Fortunately, people in the world filter the vast amount of medical information and share their personal stories in the perspective of their daily lives and activities. Patients around world would find that these personal stories are easier to relate with their personal medical conditions. Not only that they can now support each other as well.

The above irony case illustrates how the world is changing and the current health care systems are becoming costly dinosaurs, even more so now with the mobile technology. Internet and health social network have been around for some time, a lot of people in developed countries have been benefitted from it. With mobile technology, developing countries can also benefit from the online knowledge tank. Low-cost cell-phones are now equipped with sophisticated sensors: sound recorders, motion sensors, and digital cameras. Equipped with these sensors, instant two-way point to point communication of cell-phones can be utilized to search and share disease and treatment information, and support each other for better care. This will be the future of inclusive medical care that provides personalized, continuous, integrated, and affordable health care for everyone.

7.4 Medical Information Websites

Many countries now have sophisticated medical information websites for patients and their families and friends. These websites contains vast amount of information about diseases, conditions, symptoms, drugs, and even wellness issues. Table 7.2 shows a list of websites providing medical information. Figure 7.3 shows the list of health topics available from the MedlinePlus website, which is maintained by the US government. Detailed information on drugs, herbs, and supplements are provided for free of charge as shown in Fig. 7.4. This website also has video tutorials and web applications [e.g., Body Mass Index (BMI) calculator] for patients, their families, medical practitioners, and nurses. All these information are conveniently indexed by search engines Google, Yahoo, Bing, and Baidu.

Table 7.2 Online medical knowledgebase

Name	Web address		Types of Information/ Services Provided
MedlinePlus	http://www.nlm.nih.gov/medlineplus	U.S. Government	Diseases Drugs Training Tools
CDC	http://www.cdc.gov	U.S. Government	Diseases Statistics
WHO	http://www.who.int	Non-profit	Diseases Statistics
American Cancer Society	http://www.cancer.org/	Non-profit	Cancer Prevention Support and Treatment Research
Department of Health and Human Services	http://www.hhs.gov/ http://www.mentalhealth.gov/ http://www.flu.gov/ http://www.foodsafety.gov/	U.S. Government	Diseases Prevention Treatment
Other private for-profit web sites	http://www.drugs.com/ http://www.intelihealth.com/ http://www.webmd.com http://www.localhealth.com	Private organizations	Diseases Drugs Symptoms Tools
NHS	http://www.nhs.uk/	UK Government	
WebMd	http://www.webmd.com/ http://www.medicinenet.com/ http://www.medscape.com/ http://www.emedicinehealth.com/	Private	Diseases Drugs Symptoms Tools

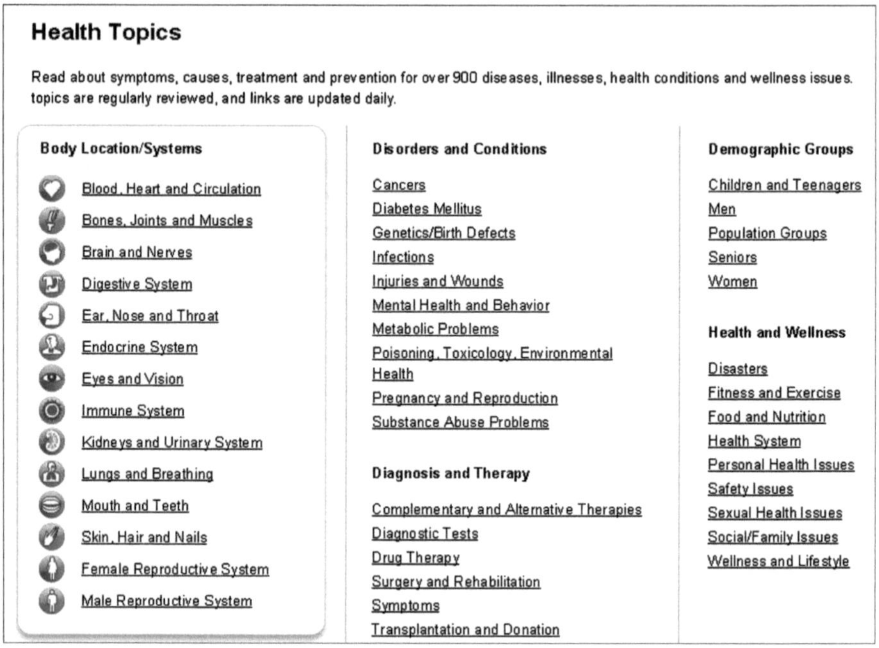

Fig. 7.3 List of health topics available from the MedlinePlus website. *Source* MedlinePlus (2010)

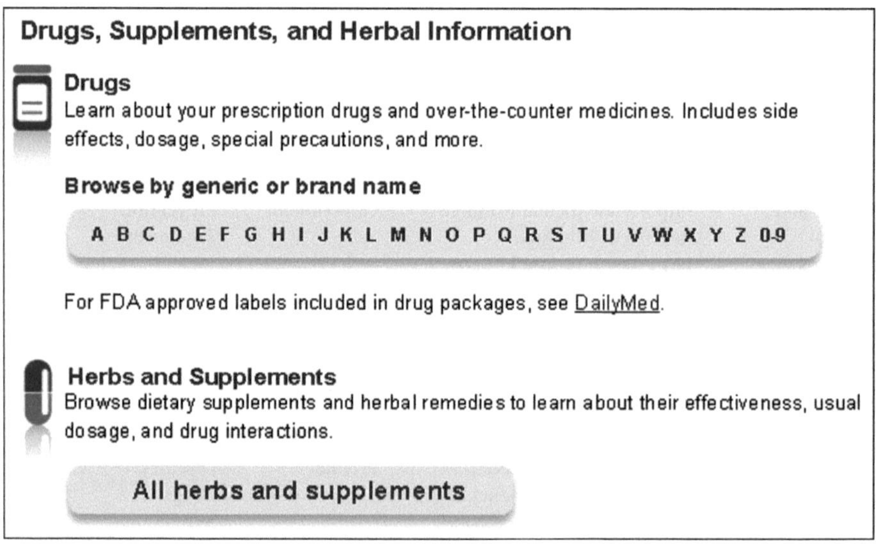

Fig. 7.4 Detailed information on drugs, herbs, and supplements are provided for free of charge from MedlinePlus website. *Source* MedlinePlus (2014)

7.5 Health Social Network

We have just realized that there are really a lot of places to explore to get information on various health conditions. But, we will soon find that there are people who are suffering from similar health problems. We will also read about their personal experiences in dealing with those problems. These encounters would make us feel relieved. After all, we are not alone suffering from similar health problems. But importantly, we will find that it is easier to find the right information that we need to deal with our health conditions. The stories are written in the context of our daily lives without medical jargons or complex drug names. We will also find that someone actually tried out some treatments, some worked and some didn't work. In addition, we will find the best solutions that really worked for people who were suffering from similar conditions. We may also find that we can actually connect with those people and share our stories as well.

Patients are starting to contribute knowledge back to the world, providing crucial information on treatments that actually worked, deriving innovation in health care. The age old doctor-to-patient centric tradition will start to disappear. There are now many tools available for anyone to create online contents to share with people. These tools are called social media tools. Table 7.3 shows a list of social media tools that people can use to create online contents.

Health social network has shown great potential to improve efficiency of diagnosis though rich information and effectiveness through continued support using communities. One important problem of the current use of social media and social network is that often sensitive information contained in assessments of health conditions cannot be shared. Another problem is finding communities or people who can support the patients. These problems can be solved by extracting a small number (less than 5) of keywords (plain words like "sport") that can be used to measure similarities of health communities (e.g., Autism parent forums) and patients. The small number of keywords can be easily shared with friends and families to find peers who might have similar conditions and provide support for each other, but without revealing any sensitive information. Figure 7.5 illustrates a

Table 7.3 Social media websites

Name	Web address	Media types
Patientslikeme	http://www.patientslikeme.com/	Online communities Stories, discussions
Autism social websites	http://www.talkaboutautism.org.uk http://www.autismweb.com/forum/ http://community.autism.org.uk/	Online communities Stories, discussions
Patient.co.uk	http://www.patient.co.uk/forums	Forums
Yahoo answers	https://answers.yahoo.com/	Forums, question and answers
WikiHow	http://www.wikihow.com/	Wiki
Webtribes	http://www.webtribes.com/	Online communities Stories, discussions

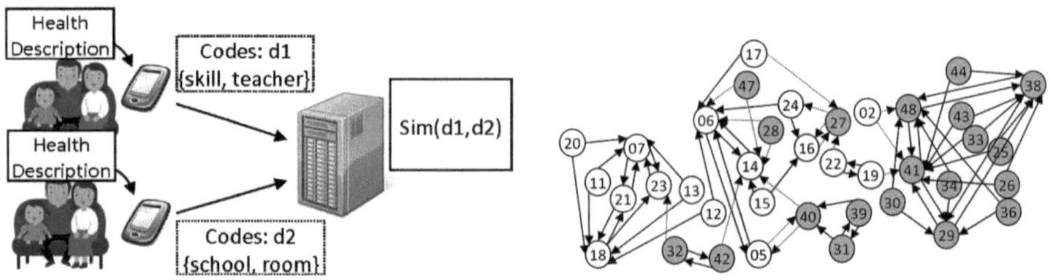

Fig. 7.5 Finding communities who can support patients using anonymous keywords. The figure *on the right* shows a result of connecting patients based on the similarity measure of the keywords. *Shaded circles* are ADHD patients. *Source* Song and Marsh (2012)

new information sharing scheme where each patient is represented as a small number of (possibly disjoint) d-words (discriminant words) and the d-words are used to measure similarities between patients without revealing sensitive personal information (Song et al. 2011; Song and Marsh 2012). The d-words are simple words like "food", and thus do not contain identifiable personal information. This makes our method an effective one-way hashing of patient assessments for a similarity measure. The d-words can be easily shared on the Internet to find peers who might have similar health conditions.

7.6 Mobile Health

Recently we have witnessed an explosion of mobile devices and mobile applications in the developed world. Thanks to the mass production of mobile devices, the cost of mobile devices has gone down dramatically. Therefore, low-cost cell phone-based medical assessment and treatment technologies have started to appear. These technologies have great potential to provide health services efficiently, timely, conveniently, and reliably, greatly improving affordability and accessibility. Examples includes mobile breathing sound analyser for diagnosing pneumonia, flue, cold and bronchitis (Lei et al. 2012, 2013, 2014; Song et al. 2011, 2013; Song and Marsh 2012) and mobile facial analysis (Song et al. 2013).

In developed countries, advanced data analysis and machine learning techniques have been successfully used for computer aided diagnosis, treatment management, and patient monitoring already shown significant potential as aids to medical and health practice (Song et al. 2011). Conventionally, specialist doctors, such as psychiatrists and general practitioners, consciously and unconsciously analyse observable symptoms, such as the language of their patients for assessment purposes using diagnostic classification systems, such as Mental Health Gap Action Programme (mhGAP) Intervention Guide (WHO 2010), BTS guidelines for the management of community acquired pneumonia (Lim et al. 2009), and DSM IV (DSM-IV-TR 2000). These conventional approaches are not only time-consuming and costly, but also may lead to subjective decisions. Recent advancements in automated assessment methods have led to solutions that overcome these limitations (Song and Marsh 2012; Lei et al. 2014).

These advanced technologies when used with low-cost cell phones now offer many solutions for reducing the cost of health services and increasing accessibility (Diederich and Song 2014). The extensive penetration of mobile phones and mobile network infrastructure allows advanced medical assessment and treatment services to reach areas that are conventionally not accessible by health care and aid workers. The available standardized interfaces, including text, image and sound recording interfaces, can be leveraged to create powerful medical assessment and treatment systems. Even low-cost cell phones are equipped with digital cameras for capturing useful information about general wellbeing states of individuals (Song et al. 2013). This rapid proliferation of mobile phones now enables us to collect high quality image data conveniently at home, even by parents of children.

Figure 7.6 illustrates the use of low-cost cell-phone for diagnosing respiratory health problems of children (Lei et al. 2013, 2014). Using basic cell phone and current Internet infrastructure, time-location tagged children's breathing sound data can be collected and stored using a web server. Its temporal and geographical patterns associated with infectious diseases can then be analysed to provide accurate, non-invasive diagnostics and treatment for lower respiratory infections in children. The clinical trial of this method showed that it is cost effective. It costs 182 times less than a General Practitioner (GP) consultation in Bangladesh (based on the cost of a mobile phone MMS at US Cents 2.5 per diagnosis, compared to US $4.55 for a GP. It provides rapid response: 1,800 times faster, than a General Practitioner (GP) consultation in Bangladesh (based on the MMS/SMS response rate of 2 min per diagnosis, compared to 3 h for a GP). The method achieved 93 % accuracy in breathing rate detection and is 77 % precision, 71 % sensitivity, and 99.5 % negative prediction value (NPV) in diagnosing lower-respiratory infections from 750 breathing sounds during our field trials. This is more accurate than the clinical examinations of GPs for pneumonia.

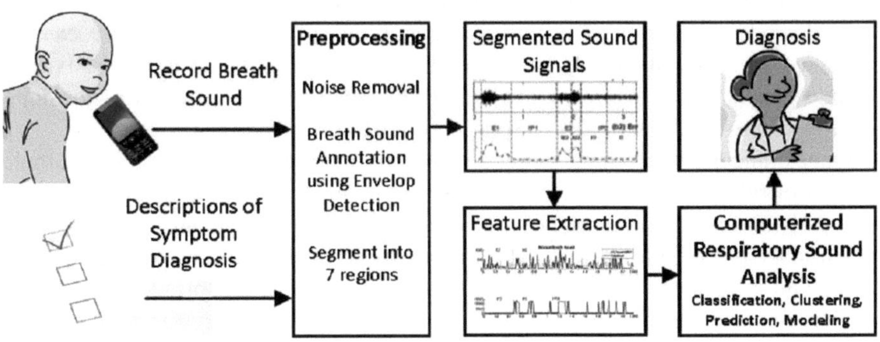

Fig. 7.6 Low-cost cell-phone based early child health intervention using children's crying and breathing sound

7.7 Other Automated Health Care Services

Telemedicine is the use of communication technologies, such as telephone, video conferencing, or the Internet, to provide and support health care to remote regions. Telemedicine was initially developed to provide health care to rural and underserved populations (Angaran 1999), but distance is no longer the major factor that defines the term. In fact, its applications have been extended to all types of health care including psychiatry consultations in inner cities (Hilty et al. 2002, 2004; Yellowlees et al. 2008). While the greatest benefits come from its use in rural and underserved populations, as it makes otherwise unavailable health care available to many patients, its driving force is now more on making health care more cost-effective, affordable, accessible, efficient, and convenient for both health care providers and consumers.

While telemedicine uses communication technologies to connect doctors and patients in remote regions, many other automated assisted technologies start to appear. These are not completely automated systems, but rather, used to assist medical practitioners for reliability, efficiency, and effectiveness. One popular approach is using computers to analyse digital images such as face images. Examples includes monitoring behaviours and diagnosing poor health in the elderly (Nambu et al. 2005), detecting emotional cues from facial expressions (Cowie et al. 2001), detection of action units (AU) (Valstar and Pantic 2011) defined by Ekman et al. (1998). Another popular approach is using computers to analyse bio-signals such as skin conductance, EEG, EMG, and EGA measurements. Examples include detection of emotional responses from galvanic skin responses (Tarvainen et al. 2001), detecting early psychological disorders using EEG and EDA signals (Frantzidis et al. 2010).

These signal analysis approaches become more sophisticated with help of advanced machine learning algorithms. Statistical analysis is an earlier approach that measures statistics of some measurements, for example the frequency of words in spoken or written content. Latent Semantic Analysis (LSA) based analysis approaches go further into analysis of hidden relations between factors. Machine learning approaches use derived features from both statistics and relations to build models of underlying health conditions. Medical practitioners often formulate clinical models in order to explain the conditions of patients. Now we can use machines to automatically generate clinical models as well. Recent advances include generation of rules for explaining diagnosis results in terms of the input features (Song and Marsh 2012; Song et al. 2011).

Once clinical models are generated by practitioners and computers, for example, as a set of rules, called knowledgebase, someone need to use the rules to diagnose new patients or explain conditions of existing patients. These are expert systems which make use of rules defined by human experts or data mining tools. Examples include an integrated medical record management system that provides a series of applications and personalized services for patients with anxiety disorder by managing patient data collection and automatically assessing stress level based on user modelling to assist personalized treatments (Panagiotakopoulos et al. 2010).

7.8 Conclusion

In this chapter we have defined IMC standard covering activities that need to be included. We illustrated that how online knowledgebase and social media are changing the landscape of medical care and driving innovation in medical care. Patients filter out information activity and propose the best methods based on their personal experiences. Publicly shared opinions are refined through online discussions and forums in collaboration with practitioners and patients. Smart technologies are operating 24/7 to provide real-time diagnosis of all kinds of health conditions such as pneumonia, flu, cold, bronchitis, and mental health problems. With the help of low-cost cell-phone, accessibility to such smart technologies is more affordable than ever before. The future has been laid out and we will soon see more IMC care systems popping out all over the world.

References

Angaran DM (1999) Telemedicine and telepharmacy: current status and future implications. Am J Health Syst Pharm 56(14):1405–1426

AAFP, AAP, ACP, AOA (2011) Guidelines for patient-centered medical home (PCMH) recognition and accreditation programs

Badlani D (2012) Africa: opportunities and challenges in a growing economy. http://www.roycefunds.com/news/global/2012/0125-africa-opportunities-challenges-growing-economy.asp

Collaborative P-CPC (2007) Joint principles of the patient centered medical home. Retrieved 15 July 2012

Cowie R, Douglas-Cowie E, Tsapatsoulis N, Votsis G, Kollias S, Fellenz W, Taylor JG (2001) Emotion recognition in human-computer interaction. IEEE Signal Process Mag 18(1):32–80

DSM-IV-TR (2000) Diagnostic and statistical manual of mental disorders, text revision, 4th edn. American Psychiatric Association, Arlington

Diederich J, Song I (2014) Mental health informatics: current approaches. In: Mental health informatics. Springer, Heidelberg, pp 1–16

Ekman P, Rosenberg EL, Heller M (1998) What the face reveals. Basic and applied studies of spontaneous expression using the facial action coding system (FACS). Psychotherapies 18 (3):179–180

Frantzidis CA, Bratsas C, Klados MA, Konstantinidis E, Lithari CD, Vivas AB, Papadelis CL, Kaldoudi E, Pappas C, Bamidis PD (2010) On the classification of emotional biosignals evoked while viewing affective pictures: an integrated data-mining-based approach for healthcare applications. IEEE Trans Inf Technol Biomed 14(2):309–318

Gantz J, Reinsel D (2010) The digital universe decade-are you ready. External publication of IDC (Analyse the Future) information and data, pp 1–16

Hilty DM, Luo JS, Morache C, Marcelo DA, Nesbitt TS (2002) Telepsychiatry—an overview for psychiatrists. CNS Drugs 16(8):527–548. doi:10.2165/00023210-200216080-00003

Hilty DM, Marks SL, Urness D, Yellowlees PM, Nesbitt TS (2004) Clinical and educational telepsychiatry applications: a review. Can J Psychiat-Rev Can Psychiat 49(1):12–23

ITU World Telecommunication (2014) ITU world telecommunication/ICT indicators database. http://www.itu.int/en/ITU-D/Statistics/Pages/stat/default.aspx. Accessed 9 Sept 2014

Lei B, Song I, Rahman SA (2012) Optimal watermarking scheme for breath sound. In: The 2012 international joint conference on neural networks (IJCNN). IEEE, Washington, pp 1–6

Lei B, Song I, Rahman SA (2013) Robust and secure watermarking scheme for breath sound. J Syst Softw 86(6):1638–1649
Lei B, Rahman SA, Song I (2014) Content-based classification of breath sound with enhanced features. Neurocomputing 141:139–147
Lim WS, Baudouin S, George R, Hill A, Jamieson C, Le Jeune I, Macfarlane J, Read R, Roberts H, Levy M (2009) BTS guidelines for the management of community acquired pneumonia in adults: update 2009. Thorax 64(3):iii1–iii55
MedlinePlus (2010) Health topics. http://www.nlm.nih.gov/medlineplus/healthtopics.html
MedlinePlus (2014) Drugs, supplements, and health information. http://www.nlm.nih.gov/medlineplus/druginformation.html
Nambu M, Nakajima K, Noshiro M, Tamura T (2005) An algorithm for the automatic detection of health conditions. IEEE Eng Med Biol Mag 24(4):38–42
Panagiotakopoulos TC, Lyras DP, Livaditis M, Sgarbas KN, Anastassopoulos GC, Lymberopoulos DK (2010) A contextual data mining approach toward assisting the treatment of anxiety disorders. IEEE Trans Inf Technol Biomed 14(3):567–581
Rothkrantz N (2011) The costs and returns to medical education. University of California, Berkeley
Song I, Marsh NV (2012) Anonymous indexing of health conditions for a similarity measure. IEEE Trans Inf Technol Biomed 16(4):737–744
Song I, Dillon D, Goh TJ, Sung M (2011) A health social network recommender system. In: Agents in principle, agents in practice—14th international conference, PRIMA 2011. Lecture notes in computer science. Springer, Heidelberg, pp 361–372
Song I, Yen NY, Vong J, Diederich J, Yellowlees P (2013) Profiling bell's palsy based on House-Brackmann score. In: IEEE symposium on computational intelligence in healthcare and e-health (CICARE, 2013). IEEE, Washington, pp 1–6
Tarvainen MP, Koistinen AS, Valkonen-Korhonen M, Partanen J, Karjalainen PA (2001) Analysis of galvanic skin responses with principal components and clustering techniques. IEEE Trans Biomed Eng 48(10):1071–1079
Valstar MF, Pantic M (2011) Fully automatic recognition of the temporal phases of facial actions. IEEE Trans Syst Man Cybern Part B: Cybern 42:28–43
WHO (2010) WHO mental health gap action programme (mhGAP) intervention guide
Yellowlees P, Burke MM, Marks SL, Hilty DM, Shore JH (2008) Emergency telepsychiatry. J Telemed Telecare 14(6):277–281. doi:10.1258/jtt.2008.080419

Chapter 8
Assessing General Well-Being Using Facial Expressions

Abstract Global cell-phone ownership has surpassed over 5 billion. The proliferation of cell phones offers an unprecedented opportunity for aid organizations and governments in developing countries for providing affordable medical services for everyone. The available standardized interfaces of low-cost cell-phones allow us to create powerful medical diagnostics systems. For instance, digital cameras of cell phones now provide easy to use interfaces for capturing useful information of various medical conditions. However, photographic images often contain private and sensitive personal information in its raw form thus considered unsuitable for many available online services. Therefore, there is a need for a computational algorithm for extracting anonymous, de-identified, digital features from captured images for assessing medical conditions and general personal wellbeing. We present a de-identified feature generation method, called Gaussian Hamming Distance (GHD). We show that GHD features are significantly correlated with personal wellbeing. Its low computational complexity makes it ideal to be used with low-cost mobile devices. Its prediction power is suitable for providing a variety of online services including recommending useful health information for improving general wellbeing.

Keywords Facial · Palsy · SVM · Face · SOM · Health informatics · eHealth · Medical data analysis

8.1 Introduction

Mobile devices have been used for rapid diagnosis of various health conditions (Lei et al. 2012), recommending communities for health social networks (Song and Marsh 2012), and improving livelihood of rural areas of developing countries (Vong et al. 2012). Low-cost cell phones are now available even in remote, hard to reach areas. In Bangladesh, a villager with a sick child would normally have to travel for 3 h just to see a nearest community doctor. The villager could instead use his US$14 cell phone to capture image, video, and/or audio data, send the data via

J. Vong and I. Song, *Emerging Technologies for Emerging Markets*,
Topics in Intelligent Engineering and Informatics 11,
DOI 10.1007/978-981-287-347-7_8

Multimedia Messaging Service (MMS) to a remote diagnostics server located in somewhere in the world for rapid diagnosis. The cost of sending an MMS message is about 10 US cents in Bangladesh. The price of cell phones and network service fees will fall further, better services will added, such as 3G, which is capable of delivering much higher-quality image and video data than 2.5G cell phone networks.

The US$14 cell phone is equipped with a digital camera that can capture an image of a face of a child. The typical resolution of such devices is about 640 by 480 pixels. A face image of a person can tell a lot about the person's health conditions. For instance, Bell's palsy is well known to be associated with respiratory infections, such as pneumonia, cold and flu (Song et al. 2013; Trad et al. 2005; Völter et al. 2004), tuberculosis (Hadfield et al. 1995), and herpes simplex (Murakami et al. 1996). However, photographic images of individuals contain identifiable private and sensitive personal information in its raw form and thus not suitable for online services. Due to the sensitive privacy issues of sharing face images, there have been little previous studies in the use of facial images for assessing health conditions. New de-identified feature extraction approaches will open up new research opportunities.

The House-Brackmann Grading System (House 1985) shown in Table 8.1 is used by clinicians to manually assess medical conditions of patients suffering from Bell's Palsy. As shown in Table 8.1, the assessment is based on measuring misalignments between the left and right hand sides of a patient's face. In telecommunications industry, Hamming distance has been used for measuring the amount

Table 8.1 House-Brackmann grading system for Bell's palsy (House 1985)

Face	Grade	Characteristics
Forehead	I. Normal II. Mild dysfunction III. Moderate dysfunction IV. Moderately severe dysfunction V. Severe dysfunction VI. Total paralysis	Normal function Slight weakness to good function Noticeable slight to moderate movement Obvious weakness or disfiguring asymmetry Barely perceptible motion No movement
Eye	I. Normal II. Mild dysfunction III. Moderate dysfunction IV. Moderately severe dysfunction V. Severe dysfunction VI. Total paralysis	Normal function Complete closure with minimal effort Obvious weakness, eye closure with effort Incomplete eye closure Barely perceptible eyelid movement No movement
Mouth	I. Normal II. Mild dysfunction III. Moderate dysfunction IV. Moderately severe dysfunction V. Severe dysfunction VI. Total paralysis	Normal function Slight asymmetry or weakness of mouth movement Obvious but no disfiguring weakness Asymmetry at rest Barely perceptible mouth movement No movement

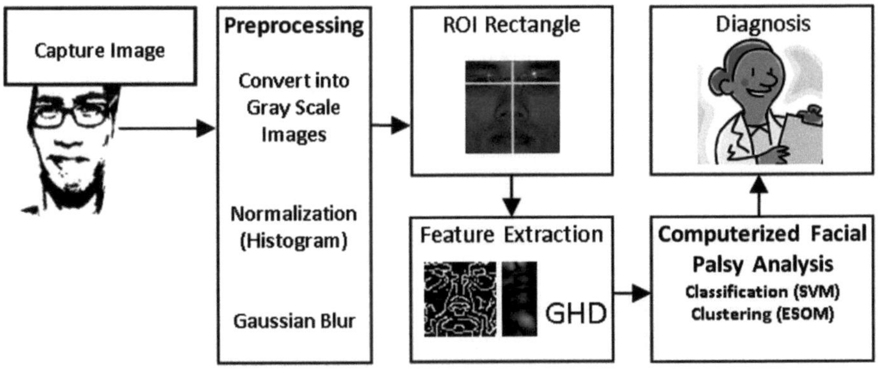

Fig. 8.1 The overall process of predicting wellbeing from facial images comprises capturing a face image, normalization, Gaussian blur, ROI detection, GHD feature generation, and prediction

of errors in a code (e.g., a string) transmitted over a telecommunication channel. In this paper, we develop a method of calculating the Hamming distance for facial images: the degree of misalignment between the left and right hand sides of a face. Unlike code or string data, misalignments of facial features can be due to some unrelated factors, such as rotation and transposition errors while taking face images.

Therefore, we have developed a computational algorithm for extracting anonymous feature data (for example, Hamming distance and Principal Components) from captured facial expression image data; and a computational algorithm [e.g., Artificial Neural Networks (ANNs) and Support Vector Machines (SVMs)] that can predict different well-being states from anonymous feature data.

In particular, we investigate how SVM is applied to automate the diagnosis of Bell's palsy from the de-identified anonymous feature data. Figure 8.1 illustrates an automated diagnosis of Bell's palsy using smart phones, such as Android mobile devices. The proliferation of mobile phones now enables us to collect high quality image data conveniently at home, even by parents of children. In particular, digital cameras of cell phones provide easy to use interfaces for capturing useful information about general well-being of individuals. Mobile devices have been used in rapid diagnosis of health conditions (Lei et al. 2012), health social networks (Song and Marsh 2012), and improving livelihood of rural areas of developing countries (Vong et al. 2012). The rapid penetration of mobile devices worldwide now opens up new opportunities to provide more convenient and cost effective means of diagnosing various health conditions associated with Bell's palsy, such as cold and flu.

8.2 Background

This section provides an overview of the core techniques focusing on telemedicine, automated health assessment technology, support vector machines (SVMs), and Emergent and Self-Organizing Map (ESOM). In particular, the significance of generating human-comprehensible explanations from classifiers, and the meanings to explain the decision-making process of a machine learning system to a human user who may not be a domain expert or have little familiarity with information technology.

8.2.1 Telemedicine

Telemedicine is the use of communication technologies, such as telephone, video conferencing, or the Internet, to provide and support health care to remote regions. Telemedicine was initially developed to provide health care to rural and underserved populations (Angaran 1999), but distance is no longer a major factor. In fact, its applications have been extended to all types of health care including psychiatry consultations in inner cities (Hilty et al. 2002, 2004; Yellowlees et al. 2008). While the greatest benefits come from its use in rural and underserved populations, its driving force is now more on making health care more cost-effective, affordable, accessible, efficient, and convenient for both health care providers and consumers. For instance, (Harrison et al. 1996) teleconferences of doctors on improving communication between primary and secondary health care, Yellowlees et al. (2008) reported case studies of telepsychiatry via videoconferencing and their potential to improve patient care and satisfaction, and reduce emergency department overcrowding. Frantzidis et al. (2010) developed a remote monitoring system for the elderly and chronically ill. Tang et al. (2011) used a multimedia system to improve medication adherence in elderly care. The growing number of elderly indicates that this form of telemedicine will become an important research area for many developed countries.

Almost all technologies reviewed herein could be used in conjunction with telemental health technology. For instance, it is possible to use automated speech analysis methods (e.g., Elvevag et al. 2009; Tilaka et al. 2013; Low et al. 2011; Song and Marsh 2012) while the patients are being interviewed by psychiatrists to provide objective analysis of speech and language disorder or to monitor improvements in mental health conditions.

Much needed research is now being devoted to the development of more sophisticated Telemedicine technologies that integrate various technologies, such as mobile breathing sound analyser (Lei et al. 2013) and recommender system (Song and Marsh 2012). The papers reviewed some of these technologies, with the aid of quantitative evaluation, rather than anecdotal qualitative evaluations (Yellowlees et al. 2008). It also examined the use of modern communication technologies for tele-health and contemporary issues in health, such as impact of technology on health services delivery, and its impact on relationship between patients and health care providers.

8.2.2 Automated Health Assessment Systems

Many health assessment methods are time consuming and highly subjective. To improve efficiency and to provide more objective assessments, various automated health assessment methods have been developed. The methods can be broadly classified based on the type of data collected and analysed. We shall now review the image analysis-based methods.

Cowie et al. (2001) reviewed the use of image analysis techniques in detecting emotional cues from facial expressions (that is, still face images) and gestures (that is, movements of facial features). According to Cowie et al. (2001), there are two emotion detection paradigms. The first paradigm detects the (seven) archetypal expressions. The second paradigm detects the non-archetypal expressions (for example, modulated, falsified, or mixed expressions). The majority of existing literature and available data clearly focuses on the method of detecting archetypal expressions (e.g., Yacoob and Devis 1996; Otsuka and Ohya 1996; Pantic and Rothkrantz 2000). Much research needs to be done for the detection of richer non-archetypal emotional expressions and gestures. These approaches now focus on the detection of action units (AU) defined by Ekman et al. (1998). For instance, Valstar and Pantic (2011) has used a facial point detector based on Gabor-feature-based boosted classifiers, achieving AU recognition rate of 95.3 %. On the application of facial expression detection method on the diagnosis of mental health problems, Stone and Wei (2011) used a facial expression detector to measure cognitive load, which is conventionally assessed using EEG or EMG measurements. However, there is little research on the use of facial emotional detection methods for clinical diagnosis of mental health problems. This may be due to the fact that image processing techniques are more difficult than other techniques, such as sound processing. In addition, large variations on emotional expressions make it difficult to develop reliable measurements that are correlated with underlying psychological problems.

Instead of relying on face detection methods, Nambu et al. (2005) developed a simple image processing technique for monitoring behaviours and diagnosing poor health in the elderly. The researchers developed an automatic diagnosis system that can assess physical conditions as well as mental conditions from behaviours of elderly watching TV. They used the recordings of the starting time of watching TV, obtained from a running monitor of the television. Initially, they tried to utilize various other sensors, such as a running monitor of electric appliances or door switches, but found the resulting data difficult for objective analysis. The data appears to be uncorrelated. The start times of watching TV are recorded by dividing each day into 15 min intervals. An interval is given a certain value, say 1, when an elderly participant starts the TV in the interval and another value, say 0, otherwise. This way, a 30 × 96 pixel image could be constructed for visualizing behaviours of the elderly watching TV. Their hypothesis was that the subjects watched the television at roughly fixed times if they were healthy. To measure productiveness of their TV watching behaviour, the researchers measured non-randomness of the patterns on the visualized behaviour using the maximum entropy method (MEM). The image compression rate was measured as an indication of the health condition. The lower compression rate (that is, the larger size of compressed image) was used as an indication of unhealthy condition.

8.2.3 Facial Palsy

Facial palsy, clinically diagnosed as facial nerve paralysis, is a facial nerve inflammation that causes facial muscle distortion. Patients who suffer from facial palsy have symptomatic, functional, communicable, and psychological problems (Terzis and Konofaos 2008). Though not fatal, chronic and severe facial palsy symptoms largely affect patients' social interaction and regular life. In order to provide accurate diagnostic information for facial palsy, studies of facial paralysis patients' database on facial expression have been undertaken. By analyzing and interpreting facial expression changes, image analysis techniques can assist in both prognosis and rehabilitation of facial nerve palsy.

One particular type of acute facial palsy, called Bell's palsy, is known to be associated with acute infectious diseases, such as pneumonia, cold and flu (Song et al. 2013; Trad et al. 2005; Völter et al. 2004), herpes simplex (Murakami et al. 1996), as well as tuberculosis (Hadfield et al. 1995). Conventionally, Bell's palsy is diagnosed using a guideline shown in Table 8.1 to assess its degree of progress. It has been shown that computerized image analysis methods utilizing techniques developed for face recognition are able to assess the degree of Bell's palsy using low resolution facial images of patients (Song et al. 2013).

8.2.4 Support Vector Machine

Cortes and Vapnik (1995) introduced support vector machines which is a novel approach to machine learning. SVMs are based on the structural risk minimization principle in order to overcome the over-fitting problems. Support vector machines find the hypotheses from the hypothesis space H of a learning system which approximately minimizes the bound on the actual error by controlling the empirical error using training samples and the complexity of the model using the VC-dimension of *H*. SVMs are universal learning systems (Joachims 1999). In their basic form, SVMs learn maximal margin hyperplanes (linear threshold functions). A hyperplane can be defined by a weight vector w and a bias *b*:

$$\mathbf{w} \cdot \mathbf{x} + b = 0 \tag{8.1}$$

The corresponding threshold function for an input vector x is then given by:

$$f(\mathbf{x}) = sign(\mathbf{w} \cdot \mathbf{x} + b) \tag{8.2}$$

However, it is possible to study polynomial classifiers, radial basis function (RBF) networks and three or more layered neural networks by mapping input data x to some other (possibly infinite dimensional) feature space $\phi(x)$ and using kernel functions $K(x_i, x_j)$ to obtain dot products, $\phi(x_i) \cdot \phi(x_j)$, of feature data.

8.2.5 Emergent Self-organizing Map

Emergent Self-Organizing Map (ESOM) (Kohonen 1990), also known as a Kohonen map having a large number of map neurons, is a technique using nonlinear projection neurons arranged in a map M, usually in 2D (i.e., $M \subset \Re^2$), for visualization of high dimensional data. Two types of ESOM grid structures have been commonly used: Hexgrid (honeycomb like) and Quadgrid (as a lattice) maps.

ESOM usually forms a (2D) grid of prototype neurons, which are represented using high-dimensional vectors with a similar number of features of input data. The density of data in the vicinity of the neurons, and the distances between the neurons, are taken into account for better visualization. For example, an ESOM map can be visualized using:

1. Distance-based Visualization called U-Map (from U-Matrix) (Ultsch 1993);
2. Density-based Visualization called P-Map (from P-Matrix) (Ultsch 2003); or
3. Distance- and Density-based Visualization called U*-Map (which combines the U and P map) (Ultsch 2003).

The different types of maps visualized the high dimensional data space on a floor space layout using a landscape like visualization for distance and/or density structure of the high dimensional data space. Structures emerge to the top of map by clustering of many neurons during the learning phase. The emerging structures are the main concept of ESOM. It can be used to achieve visualization, clustering, and classification. The clustering of ESOM is based on the U*C clustering algorithm described by Ultsch (2005).

8.3 Methodology

The project aims to develop computer algorithms that predict well-being states from facial expressions:

1. Computational algorithm of extracting anonymous feature data (e.g., Hamming distance and Principal Components) from captured facial expression image data, and

2. Prediction algorithm (e.g., Artificial Neural Networks and Support Vector Machines) that learns a function y = Model(x) of predicting different well-being states y from the anonymous feature data x. Here, Model is usually a black box.

Our main hypothesis is that an automated machine learning algorithms (e.g., ANNs) can learn a prediction function $y = Model(x)$ from a set of training sample data (X, Y), where Y = well-being states (i.e., an answer to a question in the questionnaire) and X = anonymous feature data of the corresponding facial expression image data, such as Hamming distance between left and right halves of the facial images. This can be tested by measuring accuracy, specificity, sensitivity of predicting well-being states from the anonymous features. Earlier studies (Song et al. 2011, 2013) in computational methods for the detection of facial palsy from single facial images had an accuracy of more than 70 %, suggesting that the method may be suitable for real world applications.

In order to prove the main hypothesis, we collected a total of 46 facial palsy images and 21 normal face images in our earlier studies. We included the details in this paper for reference in the later sections. We also further collected 61 face images from JCU Singapore students. The students filled in a questionnaire to describe their general well-being states, such as energetic, happy, feeling chill, feeling having a cold or flu. The questionnaire lasted about 10 min.

Each question in the questionnaire indicates a well-being state Y (e.g., Runny nose?). The facial expression images are processed to extract anonymous feature data X, such as Hamming distance. The questions in the questionnaire and features of facial expression image data form a training data set for a computational machine learning algorithm. The aim of this data set is further to test whether a computational learning algorithm (e.g., Artificial Neural Networks) can automatically generate a high-performance prediction function $y = Model(x)$.

The overall process of developing the Model function is as follows. The face images of the questionnaire are processed to generate a non-identifiable feature data set. This feature data set is then used to train machine learning algorithms (e.g., Artificial Neural Networks (ANNs), Naive Bayesian Classifier, Decision Tree, and Support Vector Machines). We then use the models to predict answers to the questions in the questionnaire (e.g., "Do you have runny Nose = Yes/No?").

We train various machine learning algorithms (Naive Bayesian Classifier, ANNs, SVM, and Decision Tree) and test their performances using the questionnaire data. Our aim is to test the computational machine learning algorithms in predicting the answers to the questions in the questionnaire from features of facial expression images.

Examples of non-identifiable features are the Hamming distances (i.e., differences) between the left and right halves of face images, edge detection and segmentation of the face images and other measures of symmetry, such as geometrical measurements of symmetry. These features cannot be used to identify individuals. We identify such features at the above second step in order to find non-identifiable features that maximize the prediction function. We can then be able to collect such non-identifiable data through various online survey methods to test further how effective the features are in predicting well-being states as patients will be more conformable to share non-identifiable data.

8.3.1 Extracting De-identified Features

Figure 8.1 illustrates the process of extracting de-identified features from facial expressions and predicting well-being states of the users as House-Brackmann Grading (see Table 8.1 for the descriptions of the grading scale) or questionnaire answers. To remove noises, a 2D Gaussian blur filter is applied on the sample face images. The images are then converted into grey scale images and normalized using a Histogram equalization process. An automated face-area detector is developed using the Haar Cascade classifier (Viola and Jones 2001) implemented on OpenCV (Open Source Computer Vision: http://www.opencv.org) API. Haar Cascade classifier combines Adaboost algorithm for feature selection and a cascade of classifiers for rapid scanning of face images. The face-area detector automatically locates a face in a sample image and places a region of interest (ROI) rectangle. The detector is calibrated such that it detects an ROI rectangle area that encloses the far edges of the eyes, the top of the eye brows, and the bottom of the lower lip. The face edges are extracted using a Sobel edge detector. The differences between left and right half images of faces are measured using Hamming Distance (Hamming 1950), i.e., the number of positions at which the corresponding pixels are different. Therefore, each image is represented by 100 Hamming distance values. Principal component analysis (PCA) is then performed on the Hamming distance features and the top-ten principal components are selected as the set of features of sample data sets. To compute the principal components, the Hamming distance features of face images are represented on a 100×67 matrix (100 rows and 67 columns), where each column is the Hamming distance features of a face image. Eigenvectors are computed using OpenCV API. The top ten eigenvectors are selected by sorting them in descending order of their eigenvalues. The images are then classified using 5-ary classifiers (a multi-class classifier).

For the training data sets, the rectangle is manually rotated and/or moved to centre the faces inside the rectangle such that the centre line of the rectangle passes through the centre of the face, that is, between the eyes and the centre of nose. The rectangle regions of the face images are then extracted, rescaled, and down sampled to 100×100 pixel square grey-scale images.

8.4 Evaluation of The De-identified Features

In our previous study (Song et al. 2011, 2013), we have collected 46 facial palsy images from hospitals and 21 normal face images from a public database. The resolutions of the facial palsy images and normal face images were about

691 × 1,024 and 150 × 180, respectively. The facial palsy images were obtained from Mater Misericordiae Health Services in Brisbane. These images show facial weaknesses, such as not being able to close both eyes, twitching of the corner of mouth when smiling, and asymmetry cheeks. The normal images were obtained from Asian Emotion Database (Chen et al. 2007). The data were then arranged into train and test data sets separately. We first graded the images manually into five scales using the House-Brackmann grades. Due to the insufficient number of images, we were not able to obtain grade VI images. We then further collected 61 smiling face images from students at the JCU Singapore campus.

Figure 8.2 shows the intermediate images of normal face and palsy images with asymmetry mouth. The right column of Fig. 8.2 shows the differences between left and right faces when we fold the face images symmetrically. The differences are measured using Hamming Distance (Hamming 1950), that is, the number of positions at which the corresponding pixels are different. Therefore, each image is represented by 100 Hamming distance values. Figure 8.3 shows all of the Hamming Distance features of the 61 students.

For our initial experiment, principal component analysis (PCA) is performed on the Hamming distance features of the 46 facial palsy images and 21 normal face images, and the top-ten principal components are selected as another set of features of the sample data sets. To compute the principal components, the Hamming distance features of face images are represented in a 100 × 67 matrix (100 rows and 67 columns), where each column is the Hamming distance features of a face image. The 5 groups of images are then classified using 5-ary classifiers.

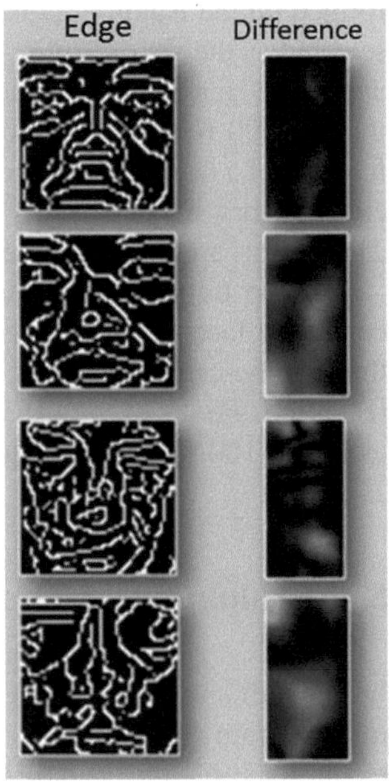

Fig. 8.2 Intermediate images of face images: significant face edges; difference between *left* and *right* face. *Source* Song et al. (2011)

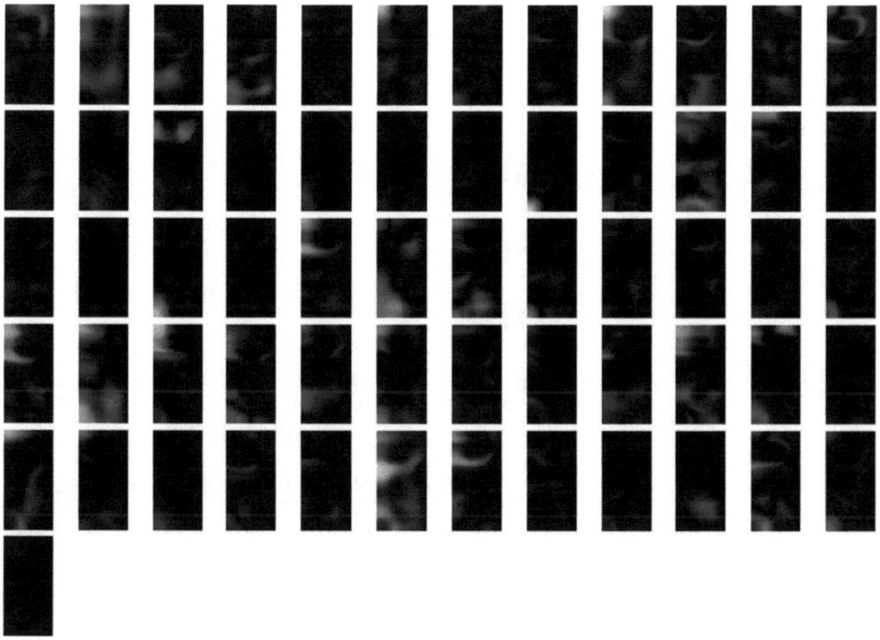

Fig. 8.3 Intermediate images of 61 smiling face images

Figure 8.4a shows the ROC (Receiver Operating Curve) curve of SVM with ten principal components, the area under the curve is 0.73. Figure 8.4b shows the ROC curve of Hamming distance with SVM which gives better results of about 0.87. It provides good curves for palsy data as shown. Table 8.2 shows the results of using ESOM classifier on the principal component features. The results are encouraging; the F-measure is above 72 %. Especially in the case of Grade II, III, and IV, the results are above 90 %.

In our second experiment, we directly use the Hamming distance features after normalizing down sampled smiling face images (50 × 50 pixels), which is much smaller than our earlier samples. The purpose of using such smaller image size to test if images transferred over Multimedia Messaging Service (MMS) can be used for diagnosis. This is often desirable in many areas where mobile broadband services, such as 3G, are not available. The features are then used to evaluate their diagnostic power. Four classifier are evaluate using leave-one-out cross validation. The classification performances of the classifiers for predicting 'feel-fit' are 0.54, 0.72, 0.74, and 0.74, respectively. The classification performances of the classifiers are 0.67, 0.75, 0.79, and 0.79, respectively. Except for Naive Bayesian, most of the classifiers performed well in predicting general well-being states. In particular, SVM and Decision tree classifiers showed good performance.

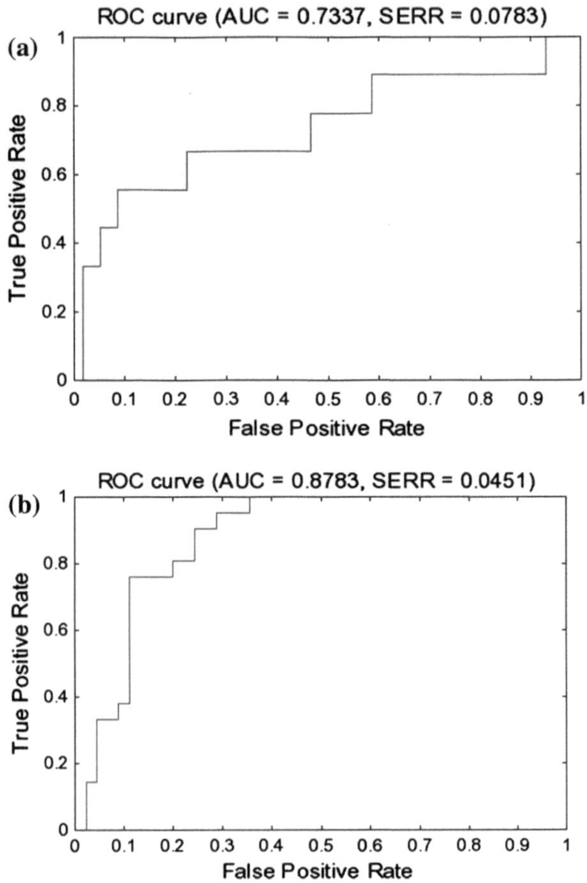

Fig. 8.4 ROC using. **a** Ten principal components. **b** Hamming distances of *left face* and *right face*. *Source* Song et al. (2011)

Table 8.2 Performance metrics using emergent self-organizing map

Grade	Generalization (%)	Recall (%)	F-measure (%)	GMts (%)	GMtr (%)
I.	59.38	91.43	72.00	59.54	93.54
II.	93.75	97.14	95.42	80.23	98.37
III.	87.5	100.00	93.33	62.06	100.00
IV.	90.63	100.00	95.08	56.73	100.00
V.	68.75	97.14	80.52	69.63	95.74

8.5 Conclusion

The results indicate that the automated machine learning algorithms can learn to predict personal wellbeing as well as medical conditions (for example, House-Brackmann score of Bell's Palsy) from de-identified features of the facial expressions. Traditional diagnosis approaches involve a medical doctor taking a thorough history of patient and determines the onset of paralysis, the rate of progression and so on. The use of automated grading system greatly reduces the duration of diagnosis while increasing consistency, because Hamming Distance (HD) is proportional to the degree of misalignment of facial features and references of HD features stored to provide comparisons. The results are very encouraging.

The HD feature extraction is computationally efficient: linear to the size of the input image. It involves convolving the input image with a small 2D Gaussian kernel and subtracting two 50-by-25 grey scale images pixel by pixel. Therefore, HD algorithm is suitable for low-cost mobile phones, which have limited memory and computational power. As HD is proportional to the misalignments in facial features, users will get immediate feedback before submitting to a remote server for more detailed analysis. This can provide timely diagnosis to patients situated in not easily accessible areas, emergency situations and areas with high medical cost, and lastly for the underserved populations.

Unlike existing medical diagnostics, such as contact thermal imaging, our method utilizes existing cell phones and infrastructure to provide a robust, non-invasive, and rapid assessment method based on the simple analysis of images. The developed system can be rapidly deployed to many needed communities over the Internet. In the next 5 years, most residents in African continent will have cell phones. Our low-cost cell phone-based data collection and diagnostics systems can be used to construct a world-wide database of diagnostic images and cases to discover the causes and spread of diseases and global patterns of disease development.

Acknowledgments This work was supported by JCU Singapore Research Grant JCUS/003/2011/IS and a grant from the Bill & Melinda Gates Foundation through the Grand Challenges Explorations Initiative (Grant Number: OPP1032125).

References

AD Tilaka, J Diederich, I Song, Teoh A (2013) Automated method for diagnosing speech and language dysfunction in schizophrenia. In: Margaret Lech IS, Peter Yellowlees, and Joachim Diederich (ed) Mental Health Informatics. Studies in Computational Intelligence. Springer, Berlin

Angaran DM (1999) Telemedicine and telepharmacy: current status and future implications. Am J Health-Syst Pharm 56(14):1405–1426

Chen EYH, Chan WSC, Chan SSM, Liu KY, Chan CLW, Wong PWC, Law YW, Yip PSF (2007) A cluster analysis of the circumstances of death in suicides in Hong Kong. Suicide Life Threat Behav 37(5):576–584

Cortes C, Vapnik V (1995) Support-vector networks. Mach Learn 20(3):273–297

Cowie R, Douglas-Cowie E, Tsapatsoulis N, Votsis G, Kollias S, Fellenz W, Taylor JG (2001) Emotion recognition in human-computer interaction. IEEE Signal Process Mag 18(1):32–80

Ekman P, Rosenberg EL, Heller M (1998) What the face reveals. Basic and applied studies of spontaneous expression using the facial action coding system (FACS). Psychotherapies 18 (3):179–180

Elvevag B, Foltz PW, Rosenstein M, DeLisi LE (2009) An automated method to analyze language use in patients with schizophrenia and their first-degree relatives. J Neuro Linguist 23(3):270–284

Frantzidis CA, Bratsas C, Klados MA, Konstantinidis E, Lithari CD, Vivas AB, Papadelis CL, Kaldoudi E, Pappas C, Bamidis PD (2010) On the classification of emotional biosignals evoked while viewing affective pictures: an integrated data-mining-based approach for healthcare applications. IEEE Trans Inf Technol Biomed 14(2):309–318

Hadfield P, Shah B, Glover G (1995) Facial palsy due to tuberculosis: the value of CT. The J Laryngol Otology 109(10):1010–1012

Hamming WR (1950) Error detecting and error correcting codes. Bell Syst Tech J 29(2):147–160

Harrison R, Clayton W, Wallace P (1996) Can telemedicine be used to improve communication between primary and secondary care? Br Med J 313(7069):1377–1381

Hilty DM, Luo JS, Morache C, Marcelo DA, Nesbitt TS (2002) Telepsychiatry—an overview for psychiatrists. CNS Drugs 16(8):527–548. doi:10.2165/00023210-200216080-00003

Hilty DM, Marks SL, Urness D, Yellowlees PM, Nesbitt TS (2004) Clinical and educational telepsychiatry applications: a review. Can J Psychiatry Rev Can Psychiatry 49(1):12–23

House JWaB DE (1985) Facial nerve grading system. Otolaryngol Head Neck Surg 93:146–147

Joachims T (1999) Making large-scale support vector machine learning practical. MIT Press, Cambridge

Kohonen T (1990) The self-organizing map. Proc IEEE 78(9):1464–1480

Lei B, Song I, Rahman SA (2012) Optimal watermarking scheme for breath sound. In: The 2012 international joint conference on Neural Networks (IJCNN), 2012. IEEE, pp 1–6

Lei B, Song I, Rahman SA (2013) Robust and secure watermarking scheme for breath sound. J Syst Softw 86:1638

Low LSA, Maddage NC, Lech M, Sheeber LB, Allen NB (2011) Detection of clinical depression in adolescents' speech during family interactions. IEEE Trans Biomed Eng 58(3 part 1):574–586

Murakami S, Mizobuchi M, Nakashiro Y, Doi T, Hato N, Yanagihara N (1996) Bell palsy and herpes simplex virus: identification of viral DNA in endoneurial fluid and muscle. Ann Int Med 124 (1-Part-1):27–30

Nambu M, Nakajima K, Noshiro M, Tamura T (2005) An algorithm for the automatic detection of health conditions. Eng Med Biol Mag IEEE 24(4):38–42

Otsuka T, Ohya J (1996) Recognition of facial expressions using HMM with continuous output probabilities, pp 323–328

Pantic M, Rothkrantz LÜM (2000) Automatic analysis of facial expressions: the state of the art. IEEE Trans Pattern Anal Mach Intell 22(12):1424–1445

Song I, Dillon D, Goh TJ, Sung M (2011) A health social network recommender system. In: Agents in principle, agents in practice—14th international conference, PRIMA 2011. Lecture notes in computer science, Springer, Berlin pp 361–372

Song I, Marsh NV (2012) Anonymous indexing of health conditions for a similarity measure. IEEE Trans Inf Technol Biomed 16(4):737–744

Song I, Yen NY, Vong J, Diederich J, Yellowlees P (2013) Profiling bell's palsy based on house-brackmann score. In: IEEE symposium on computational intelligence in healthcare and e-health (CICARE) 2013, IEEE, pp 1–6

Stone RT, Wei CS (2011) Exploring the linkage between facial expression and mental workload for arithmetic tasks, pp 616–619

Tang L, Zhou X, Yu Z, Liang Y, Zhang D, Ni H (2011) MHS: a multimedia system for improving medication adherence in elderly care. IEEE Syst J 5(4):506–517

Terzis JK, Konofaos P (2008) Nerve transfers in facial palsy. Facial Plast Surg 24(02):177–193

Trad SGJ, Dormont D, Stankoff B, Bricaire F, Caumes E (2005) Nuclear bilateral Bell's palsy and ageusia associated with *Mycoplasma pneumoniae* pulmonary infection. J Med Microbiol 54:417–419

Ultsch A (1993) Self-organizing neural networks for visualization and classification, Springer, Berlin

Ultsch A (2003) Maps for the visualization of high-dimensional data spaces. In: Proceedings of workshop on self organizing maps, 2003 pp 225–230

Ultsch A (2005) Clustering with SOM: U*C. Paper presented at the WSOM

Valstar MF, Pantic M (2011) Fully automatic recognition of the temporal phases of facial actions. IEEE Trans Syst Man and Cybern Part B: Cybern 42:28

Viola P, Jones M (2001) Rapid object detection using a boosted cascade of simple features. In: Proceedings of the 2001 IEEE computer society conference on Computer vision and pattern recognition (CVPR) 2001, vol. 511, pp I-511–I-518 IEEE

Völter C, Helms J, Weissbrich B, Rieckmann P, Abele-Horn M (2004) Frequent detection of Mycoplasma pneumoniae in Bell's palsy. Eur Arch Oto-Rhino-Laryngol Head Neck 261 (7):400–404

Vong J, Fang J, Song I (2012) Delivering financial services through mobile phone technology: a pilot study on impact of mobile money service on micro–entrepreneurs in rural Cambodia. Int J Inf Syst Change Manage 6(2):177–186

Yacoob Y, Devis LS (1996) Recognizing human facial expressions from long image sequences using optical flow. IEEE Trans Pattern Anal Mach Intell 18(6):636–642

Yellowlees P, Burke MM, Marks SL, Hilty DM, Shore JH (2008) Emergency telepsychiatry. J Telemed Telecare 14(6):277–281. doi:10.1258/jtt.2008.080419

Chapter 9
Automated Diagnoses of Respiratory Health Problems Using Breathing Sounds

Abstract Respiratory conditions, such as pneumonia, cold, flu, and bronchitis, are still the leading causes of child mortality in the world. One solution for alleviating this problem is to develop affordable respiratory assessment methods using computerized respiratory-sound analysis. This chapter illustrates how computers can be used to automatically diagnose various respiratory health problems. We use an enhanced perceptual and cepstral feature set (PerCepD) for analysing breathing sounds (BSs) for identification and classification of respiratory health problems. Classification models are developed using support vector machine (SVM) and artificial neural network (ANN) to achieve automatic detection from BS data. The high detection accuracy results validate the performance of the proposed feature sets and classification models. The experimental results also demonstrate that the high accuracy of the pathological BS data can provide reliable diagnostic suggestions for breath disorders, such as flu, pneumonia and bronchitis.

Keywords Breath sound · Support vector machine · Artificial neural network

9.1 Introduction

Acute respiratory infections are still one of the leading causes of child mortality in the world. Reports state that around 28–34 % of all child mortality under the age of 5 was due to pneumonia and 95 % of the total mortality, was from developing countries (Ruuskanen et al. 2011; Singh and Aneja 2011). It is also the major cause of death in low income countries (Ruuskanen et al. 2011). Our aim is to develop a diagnostic system for pneumonia which can be used by parents of children, in particular by the underserved population. We report on an emerging technology for diagnosing respiratory health problems that can provide cost effective methods of medical diagnosis and treatment-monitoring. Due to rapid penetration of mobile devices, breath sounds (BSs) can be easily captured using any low-cost cell phones

J. Vong and I. Song, *Emerging Technologies for Emerging Markets*,
Topics in Intelligent Engineering and Informatics 11,
DOI 10.1007/978-981-287-347-7_9

and these sounds can be sent overt to a remote server for rapid diagnosis of deadly respiratory diseases to save millions of lives.

Breath sound (BS) has been widely used in diagnosing respiratory diseases, such as flu, pneumonia and bronchitis because BS contains important indicators of respiratory health and diseases. The world health organization (WHO) has defined pneumonia solely on the basis of clinical findings obtained by visual inspection and timing of the respiratory rate (Puumalainen et al. 2008). The use of BS data for diagnosis is not new. In the past, to overcome the problems of subjective auscultation-based diagnosis of respiratory problems, various automated signal processing methods have been proposed (Yadollahi and Moussavi 2010; Low et al. 2011; Azarbarzin and Moussavi 2011; Wang et al. 2011; Matos et al. 2006; Mansy et al. 2002; Mahdhaoui and Chetouani 2011; Bahoura 2009; Arjmandi et al. 2011; Godino-Llorente and Gomez-Vilda 2004; Taplidou and Hadjileontiadis 2007; Jin et al. (2011); Shin et al. 2009; Pressler et al. 2004; Ruiz Diaz et al. 2012; Comak et al. 2007; Fonseca et al. 2007). The significance of medical applications of this study has been highlighted in various computerized and automatic sound/voice detection systems. Based on various studies, it has been suggested that computerized sound analysis is a useful and effective way to provide information of a wide variety of health conditions including respiratory diseases (Yadollahi and Moussavi 2010; Low et al. 2011; Azarbarzin and Moussavi 2011; Wang et al. 2011; Matos et al. 2006; Mansy et al. 2002; Mahdhaoui and Chetouani 2011; Bahoura 2009; Arjmandi et al. 2011; Godino-Llorente and Gomez-Vilda 2004; Taplidou and Hadjileontiadis 2007; Jin et al. (2011); Shin et al. 2009; Pressler et al. 2004; Ruiz Diaz et al. 2012; Comak et al. 2007; Fonseca et al. 2007).

However, almost all computerized respiratory sound analysis methods have been based on the data collected by using special sensors attached to the body. This is not only invasive, but also requires trained personnel to properly acquire the data. Currently, to our best knowledge, there are no computerized diagnosis methods available for pneumonia using breathing sounds collected by hand-held microphones (i.e., a non-contact method). Therefore, there is a strong need for such a method: Moreover, the existing literature strongly suggests that BS collected via non-contact methods would enable early diagnosis of respiratory diseases (Gupta et al. 1996).

BS is often regarded as a band-limited or broadband noise (Gavriely 1995). A reliable, accurate, fast method for BS data classification is essential for providing treatment and diagnosis of different respiratory diseases. A recent trend is content-based classification (Wold et al. 1996; Guo and Li 2003; Lu et al. 2003; Krishnamoorthy and Kumar 2011) techniques for audio data. The content-based classification strategy (Lu et al. 2003) shows a high accuracy compared to the traditional method. Therefore, the popular content-based classification algorithm is employed for the feature extraction of our BS data. It is known that some features cannot be captured well by just one feature set. The enhanced perceptual and mel-cepstral (PerCepD) feature, which explores features such as the delta and delta-delta coefficients to capture time dynamical information, is investigated in our study.

Compared to the pattern recognition field, the classification or separation technique is a relatively new recognition technique. Some works in the literature use hybrid models to achieve optimum classification and detection performance (Wang et al. 2011; Arjmandi and Pooyan 2012). To be more effective and accurate in the detection of the BS data, popular kernel based algorithms, such as ANN and SVM, are both adopted in the classification model to achieve higher accuracy and variability for solving practical problems (Zhang et al. 2006; Sazonov et al. 2010).

In this chapter, the hybrid feature set and content-based classification model is proposed for automatic detection of the pathological BS data collected by hand-held microphones. The main contributions of this chapter is to provide (1) an analysis of BS collected at the mouth, (2) acoustic feature extraction methods for BS so that the diagnosis of major respiratory disease can be achieved with diagnostic feature selection and (3) classification methods from BS. To the best of our knowledge, this is the first report on a method for diagnosing major respiratory diseases using BS data collected by microphones held at the mouth without any direct contact with patients and experimented with the set of features named PerCepD. Section 9.2 of this chapter introduces the methodology of feature extraction in detail. The pattern classification algorithms are presented in Sect. 9.3. Section 9.4 demonstrates the proposed method with the experimental results. Section 9.5 gives a brief discussion. Finally, conclusions and future directions are given in Sect. 9.6.

9.2 Feature Extractions

For the classification procedures, three steps are involved. Firstly, the optimal parameters are found to discriminate the BS data. Then, the reference model to evaluate the similarity of normal and pathological BS is designed and implements it practically. Lastly, the BS data is classified and recognized statistically or directly with the developed model. Figure 9.1 plots a schematic diagram of a BS detection algorithm. The algorithm comprises data acquisition, signal pre-processing, feature extraction, model training and pattern classification, and final classification results. The detailed description of the procedures and steps are as follows.

9.2.1 Data Acquisition

During the data acquisition stage, the BS signals are directly collected from the participants with digital voice recorders. The captured BS signals are usually different from centre to centre depending on data collection methods and locations, the available devices, the established practice, the training level and the particular applications. The collected BS data are from a variety of different sources with different formats. The recorded BS is re-sampled to 8,000 Hz with 16 bit resolution to facilitate the analysis techniques.

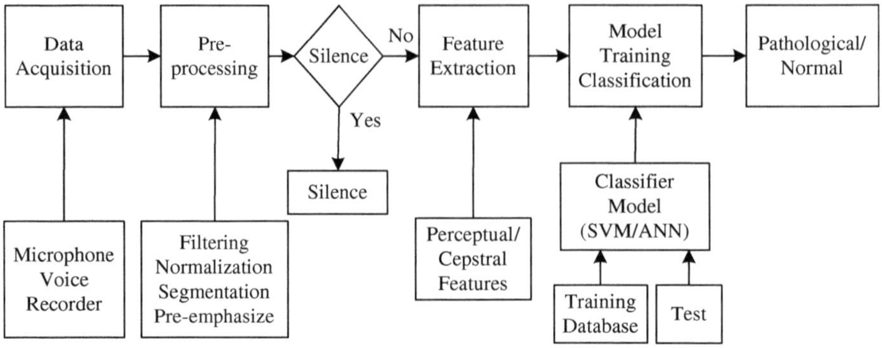

Fig. 9.1 Schematic diagram of the BS detection algorithm

9.2.2 Pre-processing

In this stage, the BS signals are prepared for smooth playback, followed by filtering and segmentation. The pre-processing of the BS includes the following steps:

Filtering:
BS is a band-limited signal (Gavriely 1995). It is filtered using a band-pass filter (9-order Butterworth filter bank) to remove the unnecessary and unwanted low-frequency and high-frequency components, and minimize movement artefacts in very low frequency and noises in high frequency.

Pre-emphasize:
The temporal feature of the BS data should be taken into consideration if the contiguous frames are overlapped. Increment of the spectrum energy in the high frequency part is done to low amplitude of the components of BS.

Normalization:
Different sampling rates and data acquisition locations will lead to signal variations, and, thus, affect the classification performance. To reduce this effect, the BS signals are normalized.

Segmentation and Windowing:
The normalized BS data are then segmented into overlapping frames with fixed length.

9.2.3 Audio Feature Extraction

After pre-processing of the BS data, the followed procedure is used to convert the BS data into a parametric representation. The enhanced perceptual and cepstral features consist of more features, such as MFCC derivatives, to further improve detection accuracy. Each feature used in our study is described below:

1. **Frequency centroid**:
   Let $S_i(w)$ represent the short-time Fourier transform of i-th frame, the frequency centroid, $C(i)$, as:

$$C(i) = \frac{\int_{L_i}^{H_i} w|S_i(w)|^2 dw}{\int_{L_i}^{H_i} |S_i(w)|^2 dw}. \tag{6}$$

2. **Bandwidth**:
   Bandwidth is defined as the square of the power of weighted average of the squared difference between the centroid and spectral components.
3. **Spectrum power**:
   The logarithm of the total spectrum power is used.
4. **Sub-band power**:
   The frequency spectrum is cut into four sub-bands with intervals. The logarithmic sub-band power is computed.
5. **Pitch**:
   The pitch frequency is detected with a pitch detection approach by tracking the peak of the normalized auto-correlation function. If the peak value is higher than an empirically chosen or specified threshold (e.g. 0.9), the pitch frequency is detected; otherwise, the frame is denoted as being non-pitched.
6. **Spectral flux**:
   Spectral flux is denoted as the spectrum variation value between the two neighbouring frames in a short time window. The timbre of a signal and the change of spectral energy distribution of two successive windows are determined by spectral flux.
7. **Spectral entropy**:
   Spectral entropy is defined as estimation of signal predictability and regularity when it indeed computes how sinusoidal the signal is. Spectral entropy allows separation of the contributions from different frequency ranges.
8. **Spectral roll-off**:
   The spectral roll-off point represents the fraction bins in the power spectrum at which 85 % of the power is at lower frequencies. It is used to separate voiced speech and music from unvoiced ones.
9. **Spectral centroid**:
   The spectral centroid is a measure of the "centre of mass" of the power spectrum. Spectral centroid measures the brightness of a sound. The higher the centroid is, the brighter the sound is.
10. **MFCC and MFCC derivatives** (**Δ and Δ-Δ**):
    MFCC (Rabiner and Juang 1993) can be estimated by using a parametric approach derived from linear prediction coefficients (LPC), or by using a

nonparametric fast fourier transform (FFT)-based approach. Time derivatives are added to the basic feature vector to incorporate the ongoing changes over multiple frames.

9.2.4 Feature Templates Formation

In our experiment, the above mentioned perceptual and cepstral features are computed over the non-silent feature. Simple statistics are computed based on the feature vectors. In our algorithm, each MFCC, Δ and Δ-Δ is an L-dimensional vector, where both the mean and standard deviation are added into the feature vector. There are 2L-dimensional vectors for MFCC, Δ and Δ-Δ respectively, while other features have one mean and one standard deviation in the feature set. Additionally, the silent ratio (number of silent frames/total frames) is regarded as one feature and combined to form the final feature vectors.

After the formation of feature vectors, the normalization is conducted. The purpose of the normalization is to remove the correlation between different features.

9.3 Classification Techniques

For the second stage, the SVM and ANN classifiers are applied for the BS classification. The BS data are identified as normal or pathological respectively with a decision rule.

9.3.1 SVM

Proposed by Vapnik (1998), SVM is a widely used classification algorithm to ensure that the local optimal solution is the global optimal solution using the statistical learning theory by the structure risk minimization principle (Low et al. 2011). The strength of SVM is that it can handle large dimensional data and overcomes the over-fitting problem. Therefore, SVM is used to address the BS classification problem in our approach.

The SVM method with sequential minimal optimization (SSVM) proposed by Platt (1998) is a very popular and typical SVM method. LIBSVM (LSVM) proposed by Hsu et al. (Chang and Lin 2011) runs faster than SSVM, as it uses the LIBSVM to build the SVM classifier. These two kernels are used since these kernels showed good result on big datasets of different respiratory sounds both in Sazonov et al. (2010) and Sun and Moore (2011). WEKA package is used to train the classifiers with default parameters.

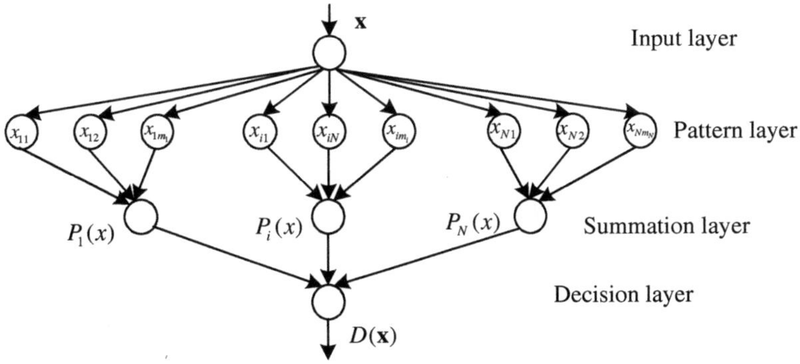

Fig. 9.2 The architecture of ANN

9.3.2 ANN

Because of the massively parallel and distributed structure, ANN (Bishop 1995) has great capabilities for learning, generalization and nonlinear approximation, and, thus, is suitable for classification. The ANN architecture is illustrated in Fig. 9.2 composed of many interconnected neurons organized in successive layers.

The summation layer uses summation units to memorize the class conditional pdfs of each class through a combination of Gaussian densities. The decision layer unit classifies the pattern x in accordance with the Bayesian decision rule based on the output of all summation layer neurons. The perceptron transforms the output according to a nonlinear activation function.

9.4 Experiments and Results

9.4.1 Database

The BS data are captured by microphones used in free field at a set distance from the subject's mouth or nose. To verify our proposed method, different experiments are carried out using our self-collected BS data recorded from people with breath disorders and people in good health without any breath disorders. Our dataset comprises a total of 90 BS data, of which 40 are normal BS data (the control subjects), and the rest are the pathological data. All the subjects were exposed in a comfortable position and asked to breathe naturally in a relaxed posture during each experiment. Written consent was obtained from each participant or from their parents (if the participant was a child). For each subject, the normal BS data were recorded for about 1 min. All the data were collected inside a room in real environment (i.e., room environment is not controlled). The recordings were uploaded to a personal computer and then converted to WAV format. Our final goal was to design a classification and detection scheme insensitive to parameters such as age, weight, gender, location and recording condition.

9.4.2 Classification Performance Evaluation

For each classifier, we used the leave-one-out (LOO) splitting method and then reported the result as average of each run. The performance of the proposed BS classification system is evaluated by quantitatively expressed detection metrics, such as average accuracy, sensitivity and specificity. Sensitivity is the true positive rate while specificity is the true negative rate. Accuracy is the proportion of correct predictions.

9.4.3 Effects of Different Classifier

Figure 9.3 shows the detection accuracy of pathological and normal BS data with different classifiers as input. In general, it can be observed that the LSVM classifier obtains the best performance among the selected classifiers, while the KNN methods achieve the lowest performance. In all cases, the detection results of the SVM and ANN classifiers are very good.

As can be seen from Fig. 9.3, the content-based classification model with supervised machine learning methods can achieve good performance with our enhanced hybrid feature dataset, because the detection accuracy results are very high. It can also be noted that LSVM and LVQ achieve the highest accuracy, while KNN obtains the lowest accuracy results in the test. Based on these findings, it can be suggested that the content-based classification by the supervised machine

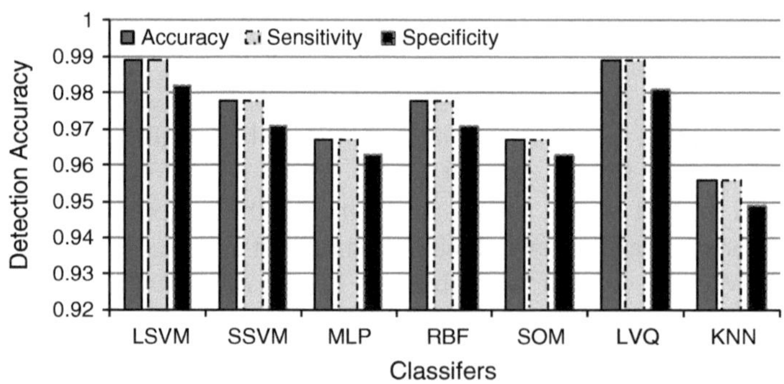

Fig. 9.3 Classification results of different classifiers. *Source* Lei et al. (2014)

learning methods is very suitable for BS data classification and detection. The SVM and ANN supervised classification methods are highly recommended for classification and detection of our BS data.

9.4.4 Experiment on Breath Disorder Detection

It is well-established that one of the main advantages of the PerCepD feature is its ability to consider simultaneously the overall information of the BS signal. This fact is important in the field of the BS disorder identification of such diseases as pneumonia, bronchitis and flu because of the noise-like effect of pathological factors.

Figure 9.4 shows that breathe disorder problem pneumonia can be effectively and successfully identified and detected by different feature sets and classifiers. It can be also noted that all the classifiers can separate pneumonia, bronchitis and flu successfully using the Cep, PerCep and PerCepD feature sets. Meanwhile, it is observed that the PerCepD feature set with the SVM classifiers, such as LSVM and SSVM, provides a higher accuracy than other feature sets as to breathing disease detection, while using other feature sets, such as Cep and PerCep, achieves less satisfactory detection accuracy. A preliminary explanation of this could be that the hybrid and effective feature sets are suitable for detecting the disordered breath diseases such as pneumonia, bronchitis and flu. It is found that the PerCepD feature set obtained a high and stable detection ratio with different classifiers for the classification of the breathing problems. Overall, the classification results for the pneumonia, bronchitis and flu indicate that the BS is detectable with the supervised classification model.

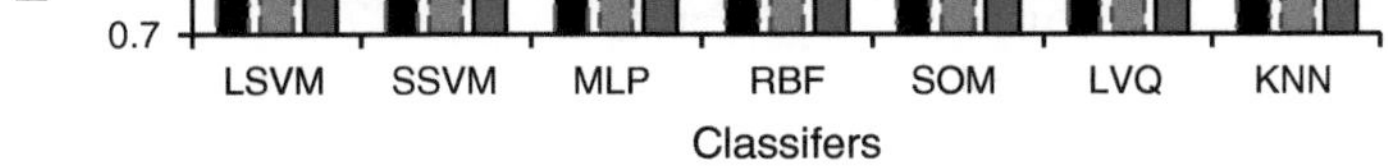

Fig. 9.4 Classification results of pneumonia diseases. *Source* Lei et al. (2014)

9.5 Discussion

Table 9.1 shows some of the best performing algorithms of speech and respiratory sounds, such as snoring, coughing, crying, and breathing sounds. It illustrates the current state-of-art in detecting various abnormal sounds for medical diagnosis. Although the datasets used for testing these algorithms are different from our dataset, we can use these performance results as the current industry benchmark of sound-based automated diagnosis algorithms. When we compare these benchmark performance results with our result, we could see that our method of detecting abnormal breathing sounds for diagnosing critical respiratory diseases (the last row in Table 9.1) significantly outperforms the benchmark results and existing breathing sound detection technology (the second last row in Table 9.1). Therefore, we can conclude that our result is better than the current industry standard in detecting abnormal respiratory sounds. We should note that the breath sounds, which are used to test our methods, are collected in noisy environments without using specialized sound recording instruments. Our method also outperforms the method by Guo and Li (2003), which is similar. This may be due to the fact that we adopt the enhanced ceptstral feature set to well represent the properties of BS data.

Table 9.1 Algorithm comparisons

References	Signal	Feature	Classifier	Accuracy
Low et al. (2011)	Speech	TEO	SVM	0.81–0.88
Azarbarzin and Moussavi (2011)	Snore sound	Suband energy	Fuzzy C-means	0.959
Matos et al. (2006)	Cough sound	MFCC	HMM	0.82
Godino-Llorente and Gomez-Vilda (2004)	Voice	MFCC	MLP/LVQ	0.95
Jin et al. (2011)	Respiratory sound	Temporal-spectral	KNN	0.924 ± 0.029
Ruiz Diaz et al. (2012)	Infant cry	Quantative feature	Template match	0.905
Comak et al. (2007)	Heart sound	Wavelet/ STFT	SVM/ANN	0.959/0.945
Fonseca et al. (2007)	Voice	Wavelet/LPC	SVM	0.9167
Guo and Li (2003)	Audio	PerCep	SVM	0.85
Lu et al. (2003)	Audio	PerCep	SVM	0.90
Dhanalakshmi et al. (2009)	Audio	MFCC	SVM/RBF	0.921/0.937
Godino-Llorente et al. (2006)	Speech recordings	MFCC	GMM	0.94 ± 0.032
Ruinskiy and Lavner (2007)	Breath sound	MFCC	Template match	0.9277
Ours	Breath sound	PerCepD	SVM/ANN	0.989

Source Lei et al. (2014)

From this table, we can observe that MFCC feature or cepstral feature is the most popular feature for sound related signal detection. Among the classification methods for the different signals and features, SVM is the most popular classifier and it provides the highest accuracy in the sense that it can be an accurate and automatic tool for signal classification.

9.6 Conclusion and Future Directions

By the measurement and analysis of the content-based classification of the recorded BS data with the supervised methods, such as SVM and ANN, the pathological BS data can be detected with very high accuracy. The SVM and ANN classifiers are evaluated and outperform the traditional KNN classifier in the detection of pathological BS data. The enhanced hybrid feature set provides more features and characteristics of the pathological BS to improve the classification accuracy. The experimental results demonstrate that the PerCepD feature of BS signal is very suitable for pathological BS detection due to its high detection accuracy. From the reported results, it is verified that BS is detectable in most cases for the medical application. This meaningful and automatic detection results can be used to provide preliminary suggestions for clinical diagnosis. This suggests that the automatic detection of pathological BS data will play a significant role in sound applications, especially in the medical field.

Acknowledgements This project is funded by a grant from the Bill & Melinda Gates Foundation through the Grand Challenges Explorations Initiative (Grant number: OPP1032125).

References

Arjmandi MK, Pooyan M (2012) An optimum algorithm in pathological voice quality assessment using wavelet-packet-based features, linear discriminant analysis and support vector machine. Biomed Signal Process Control 7(1):3–19

Arjmandi MK, Pooyan M, Mikaili M, Vali M, Moqarehzadeh A (2011) Identification of voice disorders using long-time features and support vector machine with different feature reduction methods. J Voice 25(6):e275–e289

Azarbarzin A, Moussavi ZMK (2011) Automatic and unsupervised snore sound extraction from respiratory sound signals. IEEE Trans Biomed Eng 58(5):1156–1162

Bahoura M (2009) Pattern recognition methods applied to respiratory sounds classification into normal and wheeze classes. Comput Biol Med 39(9):824–843

Bishop C (1995) Neural networks for pattern recognition. Oxford University Press, Oxford

Chang C-C, Lin C-J (2011) LIBSVM: a library for support vector machines. ACM Trans Intell Syst Technol 2, 27:1–27

Comak E, Arslan A, Turkoglu I (2007) A decision support system based on support vector machines for diagnosis of the heart valve diseases. Comput Biol Med 37(1):21–27

Dhanalakshmi P, Palanivel S, Ramalingam V (2009) Classification of audio signals using SVM and RBFNN. Expert Syst Appl 36(3 part 2):6069–6075

Fonseca ES, Guido RC, Scalassara PR, Maciel CD, Pereira JC (2007) Wavelet time-frequency analysis and least squares support vector machines for the identification of voice disorders. Comput Biol Med 37(4):571–578

Gavriely N (1995) Breath sounds methodology. CRC Press, Florida

Godino-Llorente JI, Gomez-Vilda P (2004) Automatic detection of voice impairments by means of short-term cepstral parameters and neural network based detectors. IEEE Trans Biomed Eng 51 (2):380–384

Godino-Llorente JI, Gomez-Vilda P, Blanco-Velasco M (2006) Dimensionality reduction of a pathological voice quality assessment system based on gaussian mixture models and short-term cepstral parameters. IEEE Trans Biomed Eng 53(10):1943–1953

Guo G, Li SZ (2003) Content-based audio classification and retrieval by support vector machines. IEEE Trans Neural Networks 14(1):209–215

Gupta D, Mishra S, Chaturvedi P (1996) Fast breathing in the diagnosis of pneumonia–a reassessment. J Trop Pediatr 42(4):196–199

Jin F, Krishnan S, Sattar F (2011) Adventitious sounds identification and extraction using temporal-spectral dominance-based features. IEEE Trans Biomed Eng 58(11):3078–3087

Krishnamoorthy P, Kumar S (2011) Hierarchical audio content classification system using an optimal feature selection algorithm. Multimedia Tools Appl 54(2):415–444

Lei B, Rahman SA, Song I (2014) Content-based classification of breath sound with enhanced features. Neurocomputing 141:139–147

Low LSA, Maddage NC, Lech M, Sheeber LB, Allen NB (2011) Detection of clinical depression in adolescents' speech during family interactions. IEEE Trans Biomed Eng 58(3 part 1):574–586

Lu L, Zhang HJ, Li SZ (2003) Content-based audio classification and segmentation by using support vector machines. Multimedia Syst 8(6):482–492

Mahdhaoui A, Chetouani M (2011) Supervised and semi-supervised infant-directed speech classification for parent-infant interaction analysis. Speech Commun 53(9–10):1149–1161

Mansy HA, Royston TJ, Balk RA, Sandler RH (2002) Pneumothorax detection using computerised analysis of breath sounds. Med Biol Eng Compu 40(5):526–532

Matos S, Birring SS, Pavord ID, Evans DH (2006) Detection of cough signals in continuous audio recordings using hidden Markov models. IEEE Trans Biomed Eng 53(6):1078–1083

Platt JC (1998) Sequential minimal optimization: a fast algorithm for training support vector machines. Adv Kernel Methods Support Vector Learn 208:1–21

Pressler GA, Mansfield JP, Pasterkamp H, Wodicka GR (2004) Detection of respiratory sounds at the external ear. IEEE Trans Biomed Eng 51(12):2089–2096

Puumalainen T, Quiambao B, Abucejo-Ladesma E, Lupisan S, Heiskanen-Kosma T, Ruutu P, Lucero MG, Nohynek H, Simoes EA, Riley I (2008) Clinical case review: a method to improve identification of true clinical and radiographic pneumonia in children meeting the World Health Organization definition for pneumonia. BMC Infect Dis 8:8–16

Rabiner LR, Juang BH (1993) Fundamentals of speech recognition. Prentice-Hall, Englewood Cliffs

Ruinskiy D, Lavner Y (2007) An effective algorithm for automatic detection and exact demarcation of breath sounds in speech and song signals. IEEE Trans Audio Speech Lang Process 15(3):838–850

Ruiz Diaz MA, Reyes Garcia CA, Altamirano Robles LC, Xalteno Altamirano JE, Verduzco Mendoza A (2012) Automatic infant cry analysis for the identification of qualitative features to help opportune diagnosis. Biomed Signal Process Control 7(1):43–49

Ruuskanen O, Lahti E, Jennings LC, Murdoch DR (2011) Viral pneumonia. Lancet 377 (9773):1264–1275. doi:10.1016/s0140-6736(10)61459-6

Sazonov ES, Makeyev O, Schuckers S, Lopez-Meyer P, Melanson EL, Neuman MR (2010) Automatic detection of swallowing events by acoustical means for applications of monitoring of ingestive behavior. Biomed Eng IEEE Trans 57(3):626–633

Shin SH, Hashimoto T, Hatano S (2009) Automatic detection system for cough sounds as a symptom of abnormal health condition. IEEE Trans Inf Technol Biomed 13(4):486–493

Singh V, Aneja S (2011) Pneumonia—management in the developing world. Paediatr Respir Rev 12(1):52–59. doi:10.1016/j.prrv.2010.09.011

Sun R, Moore II E (2011) Investigating glottal parameters and teager energy operators in emotion recognition. In: Affective computing and intelligent interaction. Springer, Berlin, pp 425–434

Taplidou SA, Hadjileontiadis LJ (2007) Wheeze detection based on time-frequency analysis of breath sounds. Comput Biol Med 37(8):1073–1083

Vapnik VN (1998) Statistical learning theory. Wiley, New York

Wang X, Zhang J, Yan Y (2011) Discrimination between pathological and normal voices using GMM-SVM approach. J Voice 25(1):38–43

Wold E, Blum T, Keislar D, Wheaton J (1996) Content-based classification, search, and retrieval of audio. IEEE Multimedia 3(3):27–36

Yadollahi A, Moussavi Z (2010) Automatic breath and snore sounds classification from tracheal and ambient sounds recordings. Med Eng Phys 32(9):985–990

Zhang H, Berg AC, Maire M, Malik J (2006) SVM-KNN: Discriminative nearest neighbor classification for visual category recognition. In: IEEE computer society conference on computer vision and pattern recognition, 2006. IEEE, pp 2126–2136

Chapter 10
Securing Online Medical Data

Abstract One of the challenges in providing medical services over the network connections is maintaining authenticity and integrity of the data that are transmitted over the network and stored on remote servers. To provide critical medical services, data must be free from tempering. Medical data often also contain sensitive information, and therefore require a method to prevent unauthorized access to sensitive information of patients. In this chapter, we illustrate how the methods developed for the protection of copyright can be used to achieve this. In particular, we illustrate how a watermarking scheme can be used for protecting breath sounds transmitted over the internet. We illustrate how this method can be used to insert encrypted source and identity information in breath sounds while maintaining significant biological signals for proper diagnosis. The method can be applied to any types of data including image, video, and sound data. We show experimental results to demonstrate that the proposed watermarking scheme obtains good robustness against common manipulation attacks and preserves imperceptivity. The performance comparison results verify that the scheme outperforms existing approaches in terms of robustness and imperceptibility.

Keywords Digital watermarking · Secure medical data · Breath sound

10.1 Introduction

Advances in modern communication technologies, in particular the Internet and health social networks, allows anyone to have access to, make copies of, and distribute digital contents. However, this also causes a problem, as anybody can modify or tamper with important data, such as breathe sounds for detecting infectious diseases. Digital watermarking is one of the solutions to address this problem and is used for digital rights protection, ownership verification and security purposes (Bender et al. 1996). For medical data, such as breath sounds containing medical information, conventional watermarking is not appropriate because of the

J. Vong and I. Song, *Emerging Technologies for Emerging Markets*,
Topics in Intelligent Engineering and Informatics 11,
DOI 10.1007/978-981-287-347-7_10

signal characteristics and distortion problems. Therefore, the evidence strongly suggests that a new watermarking scheme should be designed.

Watermarking of medical signals is different from the better known audio watermarking for intellectual property protection. However, the small band width medical signals can still use the same analysis techniques because of their similar signal characteristics. Most of the current digital audio and speech watermarking schemes mainly focus on one or two transform combinations. The justification for the combination of transforms is that the combination of transforms for watermarking can achieve the optimal watermarking performance in terms of robustness and imperceptibility. We will incorporate the following modern techniques to achieve this: combining lifting wavelet transform (LWT), discrete cosine transform (DCT), singular value decomposition (SVD), dither modulation (DM) quantization, and particle swarm optimization (PSO) techniques. The reason to apply LWT, DCT and SVD sequentially is that the system can achieve optimal performance in terms of robustness and transparency in this order.

Particle swarm optimization (PSO) has been applied in the watermarking field to improve the performance as an evolutionary optimization technique (Kennedy and Eberhart (2001). PSO shares similar characteristics as genetic algorithm (GA), but it was proven that PSO could obtain better results in a faster and cheaper way. There are also fewer parameters in PSO to adjust compared with GA method. Initializing a population of random solutions first, PSO algorithm searches for optimal values by updating iterations. Therefore, PSO has been used for approaches in a wide range of fields.

This chapter illustrates the robust watermarking technique for breath sounds, which was developed by Lei et al. (2013). This method is based on a unique system in which breath sound data are recorded by digital devices. The proposed watermarking method based on LWT-DCT-SVD for breath sounds in the medical application is integrated with the PSO to improve the watermarking performance. A cat map is utilized to scramble the watermark data and increase confidentiality. The quantization steps are optimized by the PSO heuristic algorithm. The main significance of our proposed method is for the data tampering protection of our self-collected breath sounds for providing health information. Moreover, the meaningful image logo can also provide identification. The organization of this paper is as follows. Section 10.2 presents our proposed method in detail. Performance analysis is discussed in Sect. 10.3. Experimental results are presented in Sect. 10.4. Finally, Sect. 10.5 concludes the paper.

10.2 Overview of Technologies and Algorithms Used for Watermarking

10.2.1 Particle Swam Optimization (PSO)

In PSO, each particle searches for optimal coordinates in the problem space. The coordinates are related with the best solution (fitness or objective value) (Kennedy and Eberhart 2001). The objective value (called pbest) is also stored. Another "best" value (called gbest) is found by the particle swarm optimizer, which is the best value obtained by any particle in the neighbours of the particle. When a particle takes all the population as its topological neighbours, the best value is a global best. At each time step, the PSO process comprises changing and accelerating the velocity of each particle toward its pbest and gbest locations (PSO local version). Acceleration is weighted by a random term, with separate random numbers being generated for acceleration toward pbest and gbest locations.

10.2.2 Lifting Wavelet Transform (LWT)

LWT utilizes lifting wavelet in the integer domain to simplify the problem, and thus has strength over the traditional wavelet transform. There are three main steps of LWT, that is, split/merge, prediction and update, as described below.

(1) **Split step**:
The split step just splits the input signal $X(n)$ into even and odd parts.
(2) **Prediction step**:
The prediction step predicts odd samples from even samples, which is a dual lifting step. The two signal subsets from the split process should be closely correlated. The difference between the predicted value of $P[Xe(n)]$ and the real value of $Xo(n)$ is called the detail signal $d(n)$.
(3) **Update step**:
The update step introduces the update operator $\tilde{U}[\bullet]$, which adopts the detail signal $d(n)$ to update even samples. The update step is the lifting of low-pass subband with high-pass subband, and thus it is often viewed as a low-pass filter.

10.2.3 Singular Value Decomposition (SVD) Principle

SVD (Andrews and Patterson 1976) is a matrix factorization which decomposes a matrix into three matrices of the same size. For example, a real matrix I of size $N \times M$ can be decomposed into a product of 3 matrices: $I = USV^T$, where the superscript T denotes matrix transposition, U and V are real orthogonal $N \times M$ matrices with small singular values, and S is $N \times M$ size diagonal matrix with large singular values.

10.2.4 Cat Map Encryption

The chaotic sequence often has low correlation sidelobes which means good invariance to disturbance. In order to disturb the pixel space relationship of the binary watermark image, the watermark is first permuted by the chaotic map. To enhance the security of the proposed watermarking scheme, a watermark scrambling algorithm is used at first. The two-dimensional binary image is then converted into one-dimensional sequence in order to embed it into the breath sound signal. It is further encrypted using random permutation to ensure security. Finally, the cat map encrypted data, $w(n)$, is used for the watermark embedding.

10.2.5 Watermark Embedding

Figure 10.1 describes the block diagram of the watermark embedding algorithm. In our scheme, the combined features of LWT, DCT, SVD and DM are fully utilized.

We chose the popular DM quantization method in the embedding process because of its good robustness and blind nature. The breath sound is first preprocessed with the normalization method for providing data consistency and improving the watermarking performance. The following steps describe the embedding process:

Step 1: Perform LWT on the host breath sound signal, SB, to generate the coefficients of approximations and details.

Step 2: We need to embed as many watermark bits as possible in the LWT coefficients. To achieve optimal performance, DCT is only applied to the low frequency approximate coefficients of the signal.

Step 3: Cut the generated LWT-DCT coefficients into different blocks. Specifically, the block length is equal to dividing the length of the approximate coefficients by the watermark length.

Step 4: Conduct SVD transformation on each block to produce singular values.

Step 5: In order to guarantee robustness and transparency of watermarking scheme, the watermarking system inserts the watermark bits into the LWT-DCT coefficients by adaptive DM quantization method. The first singular values of each block have the highest energy values, and thus they are utilized to embed the watermark.

Step 6: Inverse SVD transformation is conducted to obtain the watermarked block.

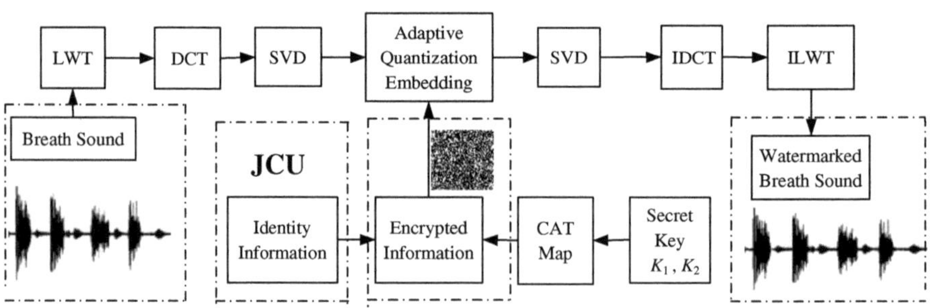

Fig. 10.1 Diagram of the watermark embedding procedure. *Source* Lei et al. (2013)

Step 7: Apply DCT to the modified signal.
Step 8: K-level inverse LWT is carried out to get the finally watermarked breath sound signal.

10.2.6 Watermark Extraction

Figure 10.2 illustrates the watermark extraction steps. The watermark extraction process is described as follows:

Step 1: Perform DWT to the watermarked and possibly distorted host breath sound signal, to obtain the LWT coefficients.
Step 2: Apply DCT on the low frequency part.
Step 3: LWT-DCT coefficient is divided into different blocks with the same block length as that in the watermark embedding process.
Step 4: SVD transformation is applied on each block to produce singular values.
Step 5: The final embedded image logo can be obtained by decrypting the signal with the same cat map in the embedding procedure.

10.2.7 Performance Optimization Using PSO

The quantization step for watermarking is signal-dependent, and thus a fixed value for all test signals is not optimized. Therefore, different host signals or watermarks require different quantization steps depending on their signal characteristics. To address this issue, the evolutionary algorithm, PSO, is used to search for the best quantization steps for different host signals and watermarks. In our scheme, PSO is applied under this constraint because the quantization step for our watermarking is in the range of [0, 1].

Signal processing attacks, such as resampling, MP3 compression, noise addition and cropping are commonly used hence they are selected to be embedded in our

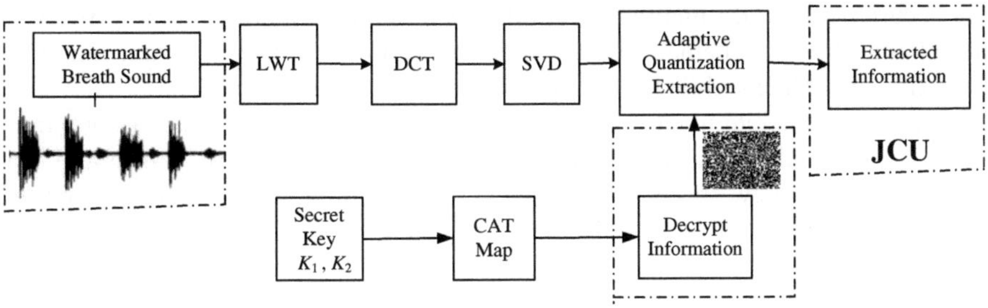

Fig. 10.2 Diagram of the watermark extraction procedure. *Source* Lei et al. (2013)

PSO training procedure to measure watermarking performance. In the watermarking field, the properties of watermark robustness, imperceptibility and capacity are mutually contradictory to each other. Since we only want to achieve a trade-off between watermark imperceptibility and robustness, the watermark capacity is fixed in the PSO process. The objective value function is designed as a function of both imperceptibility and robustness to obtain the optimal performance by PSO.

10.3 Performance Analysis

10.3.1 Watermark Capacity

The watermark capacity (also known as the payload) of a watermarking scheme is defined as the number of bits that can be embedded and recovered in the audio stream, and often measured in bits-per-seconds (bps).

10.3.2 Error Analysis

Bit Error Rate (BER) is adopted to evaluate the proposed watermarking scheme. It is used to evaluate our water marking methods against signal distortions or collusion attacks. To decide whether there is a watermark, the original watermark w is compared with the extracted watermark w'. If the BER between w and w' is less than a user-defined threshold L, there is a watermark, otherwise, there is no watermark. Threshold L is determined by the probability of the detection error due to a false alarm or rejection detection (Bhat et al. 2010). The false alarm rate of our scheme demonstrated good error probability of the proposed algorithm and it has sufficient confidence to counter on the collusion attack to decide the sign of the hidden watermark.

10.3.3 Key Space Analysis

The security of an information system depends on keys rather than the privacy of the scheme. In our scheme, keys or initial conditions are floating-point numbers. The exponent is fixed while the significant number may be varied. The two keys K_1, K_2 is used to generate a cat map, thereby enhancing the security of the proposed scheme. The total key space of our watermarking scheme is 10^{34}, meaning that there is enough key space to guarantee high confidentiality of our proposed watermarking system. Based on this key space analysis, it can be concluded that the embedded watermarks are secure to attackers who try to exhaustively or statistically detect and read them. All in all, our proposed scheme with such a long key is adequate for reliable and practical use.

10.3.4 Security Analysis of the Estimation Attack

In order to decrypt the encryption function adopted to construct the content-dependent watermark, the probability of inversing the scrambling function is of great significance, which means the single watermarked signal counterfeit attack or estimation attack. The probability of a successful attack mostly relies on the randomness of watermarking. As a result, it is more difficult to forge or estimate a watermark when the watermarking scheme has a higher degree of randomness. The randomness of watermark detection and watermarking region selection prevents an attacker from estimating the exact range and location of the watermarked area.

10.4 Experimental Results

To evaluate the performance of the watermarking scheme, we collected a total of 90 breath sounds from different children under the age of 9. 40 of them were recorded in indoor environments and 50 of them were recorded in outdoor environments. General well-being questions were asked about each participant (e.g., "whether the child is energetic or sluggish, having pneumonia, cold or flu") and the information gathered is embedded with the breath sounds to identify the well-being state of the children. The recordings were uploaded to a personal computer, converted to the WAV file format with a fixed duration time and then divided into different segments to facilitate the analysis. The sounds were sampled at 32 kHz with 16 bit resolution.

A 64×64 bit binary JCU image logo was adopted in our watermarking method. The duration time of the breath sound data was 32.768 s. In our observations, a larger decomposition level could not increase the watermarking robustness and caused intensive computation. Thus, we chose 3 as the decomposition level of the Haar wavelet as a trade-off between robustness and imperceptibility. BER was adopted to evaluate the proposed watermarking scheme.

10.4.1 Imperceptibility

Two representative breath sound samples were used in our scheme for the performance comparison. Figure 10.3 plots the time domain waveforms and frequency domain spectrogram of the original and watermarked breath sounds, respectively. No observed visual differences were found between the time domain waveforms. The low frequency part of the frequency domain spectrum of the original and watermarked signal were almost the same, while the high frequency part of the spectrogram of the original and watermarked signal had only some noise differences, which verified the successful insertion of the watermark and the good

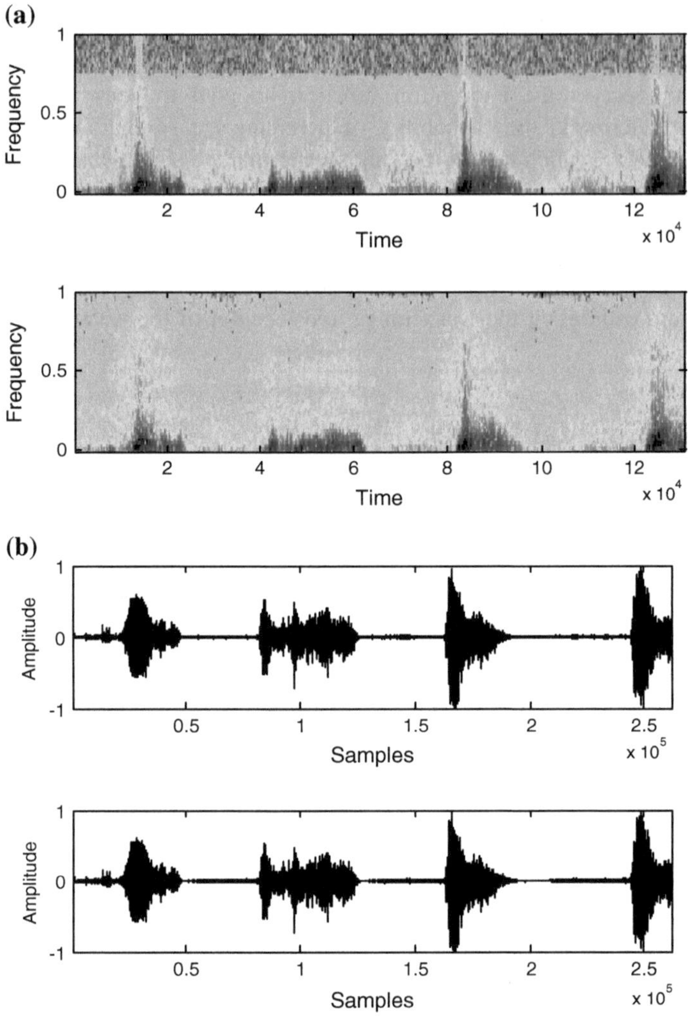

Fig. 10.3 **a** Spectrogram of the original and watermarked breath sound1; **b** Waveform of the original and watermarked breath sounds (SNR = 24.3143, LLR = 1.851). *Source* Lei et al. (2013)

imperceptibility of the watermarking scheme. Moreover, the high SNR and LLR values after watermarking also validated good transparency of the breath sound watermarking scheme objectively.

10.4.2 Results of PSO Technique

The conflicting problem of imperceptibility and robustness of the watermark was addressed by the PSO based optimization method. Figure 10.4 shows the results of

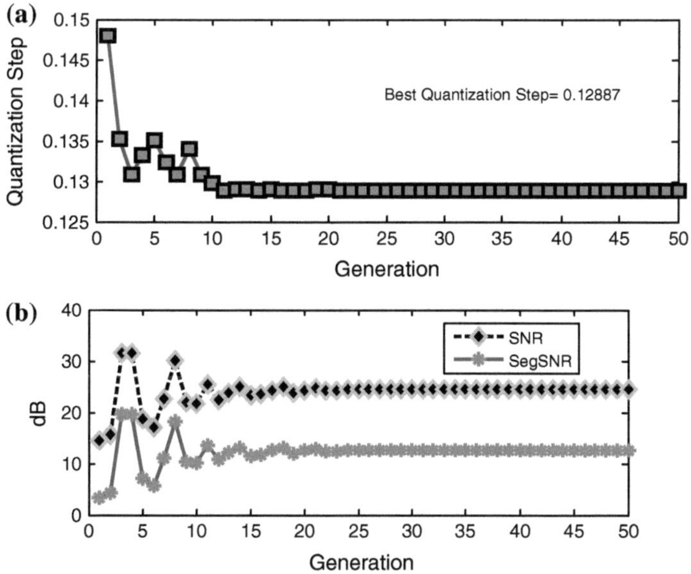

Fig. 10.4 Quantization step, SNR, SSNR, Corr and LLR values of breath sounds 1 with PSO training using 20 particles after 50 generations; **a** Best quantization step results (Best quantization step = 0.1289); **b** Best SNR and SSNR results (best SNR = 24.7532 and best SSNR = 13.2821). *Source* Lei et al. (2013)

PSO optimization of our proposed LWT-DCT-SVD watermarking method after 50 generations of the two presentive breath sound signals. The quantization steps, SNR, SSNR, LLR, and the average Corr values of the four attacks gradually converged to the saturation values as the number of generations was increased. These results validate the positive effect of PSO training and the successful optimization technique. These parameters were optimally varied to achieve the most desirable ones for host signals according to their different breath sound characteristics. Moreover, it was observed that 50 generations were enough to achieve the optimal values from these figures. Therefore, 50 generations were used in our simulations.

10.4.3 Robustness

General signal processing attacks such as resampling (resampling to half of the sampling frequency and then back to the sampling frequency), requantization (16 bits-8bits-16bits), MP3 compression (at bit rate of 112, 96 and 64 kbit/s), cropping (10 % of the original samples are cropped) and noise addition (with the SNR 30 dB) were conducted to validate the robustness of our scheme. The robustness results and extracted watermarks under various kinds of attacks are shown in Table 10.1. It can

Table 10.1 Robustness test results

Attacks	Original watermark	Resampling	Requantization
Extracted watermark	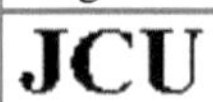	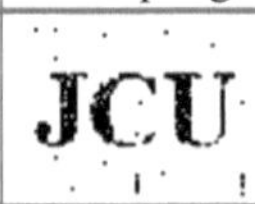	
BER (%)	0	2.54	0
Corr	1	0.9727	1
Attacks	Low pass filtering	Additive noise	Cropping
Extracted watermark			
BER (%)	0.17	0.54	0.07
Corr	0.9988	0.9962	0.9995
Attacks (kbps)	MP3 96	MP3 64	MP3 56
Extracted watermark		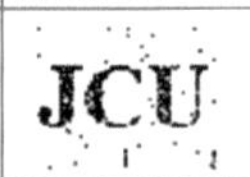	
BER (%)	0	4.03	4.81
Corr	1	0.9709	0.9647

Source Lei et al. (2013)

be noted that the watermarks under attacks are still visually recognizable hence can be used for data source identification and data tampering detection purposes for breath sound signals.

The overall robustness performance after attacks of our breath sound watermarking scheme is quite good. This watermarking scheme can achieve source identification and data tampering protection based on the robustness performance and the visually recognizable extracted watermarks after attacks.

10.5 Conclusions

In this paper, a robust and blind watermarking scheme for medical breath sounds based on LWT-DCT with the exploration of SVD properties and DM quantization was proposed. Our scheme is very robust to various common signal processing attacks. This was validated by the simulation results. In our optimization process, PSO was used to make an optimal trade-off between imperceptibility and robustness through effective selection of quantization steps to locate the best parameters to insert the watermarks. The quantization steps were optimally adapted to achieve the most suitable performance for various breath sounds with different characteristics. In addition the proposed watermarking scheme achieved very good transparency results. The experimental results revealed that our proposed method can increase the robustness of the embedded watermark against various attacks in comparison to previous work. The proposed method performed much better than the DCT method for almost all of the images and attacks. As the identification information can be exactly extracted without affecting the imperceptibility of the breath sounds, it is especially suited to medical breath sound applications.

Acknowledgments This project is funded by a grant from the Bill & Melinda Gates Foundation through the Grand Challenges Explorations Initiative (Grant Number: OPP1032125).

References

Andrews H, Patterson C (1976) Singular value decompositions and digital image processing. IEEE Trans Acoust Speech Signal Process 24(1):26–53. doi:10.1109/TASSP.1976.1162766

Bender W, Gruhl D, Morimoto N, Lu A (1996) Techniques for data hiding. IBM Syst J 35(3–4):313–336. doi:10.1147/sj.353.0313

Bhat KV, Sengupta I, Das A (2010) An adaptive audio watermarking based on the singular value decomposition in the wavelet domain. Digit Signal Process Rev J 20(6), 1547–1558

Kennedy J, Eberhart RC (2001) Swarm intelligence. Morgan Kaufmann Publishers Inc., San Francisco

Lei B, Song I, Rahman SA (2013) Robust and secure watermarking scheme for breath sound. J Syst Software 86(6):1638–1649

Part III
Education

Chapter 11
Technology in Education

Abstract Supporting diverse and rapidly changing learning styles of new digital age generations is a major hurdle in the age of massification of education markets. The institutions must now utilize the unprecedented network speed and mobile technology to stimulating learning environments for the new digital age generations. Institutions need to accommodate different learning styles of students often comprising of working professionals, different age groups and students from cultural backgrounds. Researchers have been trying solve these challenges through flexible learning and blended learning approaches. Faced with new digital age generations, unprecedented network speed and mobile technology, researchers now frequently quote mobilized curriculum, social-interaction oriented learning, personalized learning and collaborative mobile learning. This chapter reviews relevant education models, learning styles, literature, and technology to help identify new emerging research problems and research methodologies in m-learning.

Keywords Technology in education · Blended learning · Mobile learning · Collaborative learning · m-Learning

11.1 Introduction

In the age of massification of education market, educational institutions must accommodate changing student profiles and diverse student cohorts through technological and pedagogical advancements (Devlin 2012). Institutions need to accommodate different learning styles of students often comprising of working professionals, different age groups and students from cultural backgrounds. Researchers have been trying solve these challenges through flexible learning (McInnis and Hartley 2002; Devlin 2012; Lefoe and Hedberg 2006) and blended learning (Sharpe et al. 2006) approaches. Faced with new digital age generations (Conole 2004), unprecedented network speed and mobile technology, researchers now frequently quote mobilized curriculum (Zhang et al. 2010), social-interaction

J. Vong and I. Song, *Emerging Technologies for Emerging Markets*,
Topics in Intelligent Engineering and Informatics 11,
DOI 10.1007/978-981-287-347-7_11

oriented learning (Ting 2013), personalized learning (Song et al. 2012) and collaborative mobile learning (Boticki et al. 2013).

This chapter reviews relevant literature to identify research problems and research methodologies in m-learning. In Sects. 11.2–11.4, we review literature in mobile learning for mobilized curriculum, formative assessment, and enhancing social experiences. In Sect. 11.5, we summarize learning types, research methodologies, evaluation approaches, data gathering approaches, and data analysis approaches used by m-learning researchers. In Sect. 11.6, we conclude with identified gaps in literature.

11.2 Mobile Learning for Enhancing Social Experiences

Ting (2013) viewed that social interaction is a key ingredient in the designs of mobile learning practices. He proposed that mobile learning could enable students interact with peers, teachers, and subject contents more efficiently and unobtrusively. This would then provide a more harmonized environment where students could construct their learning upon both physical and social contexts augmented on the learning content. Boticki et al. (2013) proposed that mobile-learning can provide efficient collaborative learning environment to enhance in-class learning experiences. Boticki et al. (2013) proposed a mobile-technology platform for supporting collaborative learning in class and evaluated the platform in teaching two Primary 3 subjects in class: Chinese characters and mathematics. Ting (2013) proposed a curriculum design practice, called interwoven learning interactions, for creating more socially interactive learning environments.

11.3 Mobile Learning for Formative Assessment

In health care education, competency assessments are typically assessed summatively due to the unique curriculum design in this sector. In a similar reason, not much research has been done in the use of mobile learning in health care education. Furthermore, the use of mobile technology in health care education initially met with scepticism. Kneebone and Brenton (2005) reported that the usage of mobile devices did not improve the quality of assessment tasks or even reduced efficiency. Other studies suggested that special types of mobile devices were required to solve these problems (Kramer 2009). Coulby et al. (2011), however, suggested that identifying an appropriate use of mobile devices is key to solving the problems. They hypothesized that m-learning could be used to conduct formative assessment in competency assessments in health care education. This was an important observation as competency assessments were typically assessed summatively in health care education. They used mobile devices to conduct formative assessment on students enrolled in work-based placements for Mini-Clinical Evaluation Exercise (CEX), which is a competency based assessment in health care education. They conducted an initial pilot study on four final-year medical students. The researchers observed the initial face-to-face training of the students on the use of mobile devices for their training. The formative assessment activities were also observed. For triangulation of data collection, they have also conducted focused group interviews. After positive feedback from the initial pilot study, another nine final-year medical students were invited to the study. They used grounded theory approach to analyse the qualitative data they have gathered focusing on discovering concepts, relationships between concepts, and themes, in particular, attitudinal and behavioural patterns and themes associated with the use of mobile technology. They have observed that it was easy for teachers to review students' progress more frequently as they could do remotely. As a result, students could be able to receive feedback more frequently as well. Students could monitor their progress through continuous feedback and teachers could chart incremental improvements of individual students. Most students felt the formative assessment greatly helped them to pass the summative assessment tests at the end of their placements.

In a similar attempt, Hardyman et al. (2013) used mobile devices to support trainee doctors in workplace assessments. They argued that supporting the process of knowledge acquisition and practices was the main problem for training medical doctors due to the vast amount of knowledge, which were only applicable under certain specific contexts. For example, trainee doctors had to communicate with senior doctors, peers, and nursing and other medical staff, in that order, to solve complex problems, For simple information queries, they used hard copy texts and various resources. Therefore, the researchers designed iDoc (a mobile information access platform for medical knowledge) to support Foundation Year 1 and 2 trainee doctors. They have designed an evaluation on 440 trainee doctors using a descriptive framework of e-learning (Ellaway 2010) and contextual mobile learning approach (David et al. 2007). They have collected narrative case reports and responses to questionnaire for their data gathering. The questionnaire responses were analysed using SPSS quantitatively. The narrative cases reports describe trainees' accounts of the use of mobile technology during their training and the accounts were analysed qualitatively using NVivo to classify the case reports into three types of scenarios. The questionnaire data were then used to find out types of information sources, information seeking behaviours of the trainee doctors, and perceived benefits of iDoc during their training. Both the questionnaire responses and the case reports suggested that trainee doctors found iDoc useful in finding information and getting feedback from senior doctors more quickly during patient consultations. iDoc improved the overall efficiency of information access and communication during their training.

11.4 Mobilized Curriculums

Zhang et al. (2010) developed a methodology for re-designing existing curriculums into mobilized curriculums. The authors applied the methodology to redesign existing 2nd year primary-school science curriculums in Singapore. The mobilized curriculums were designed to allow students to participate in inquiry-based learning in class or out of class using their mobile devices. The authors hypothesized that mobilized curriculums, referred to as "mobilized lesson" by Norris and Soloway (2008), might be able to provide an infrastructure that fosters student centred learning due to the pervasive nature of mobile technologies. Students would have 24/7 access to the mobile devices to participate in both informal and formal learning in class or out of class. Before a pilot study, the authors conducted a field study over a six month period to identify opportunities for improving learning using mobile technologies. The pilot study was done using GoKnow Mobile Learning Environment (MLE) running on HTC Windows Mobile 6 Smartphone. The researchers and teachers co-designed 2nd year primary-school science curriculums by first deconstructing existing curriculums, then re-designing learning goals and learning activities to utilize various software tools, such as timers and calculators, available on the Smartphone and GoKnow MLE. The researchers selected five affordances of mobile technologies from previous studies done by Churchill and Churchill (2008) and Patten et al. (2006) as their guidelines for choosing the features for redesigning the curriculums: (1) Personal learning tool, (2) Contextualized learning tool, (3) Collaboration tool, (4) Data Capture tool, and (5) Visual articulation or translation tool. The pilot study was then conducted in nine Primary classes. The pilot study showed that students participated in the redesigned mobilized curriculum performed significantly better than the rest of classes in the final exams. The researchers also conducted continuous observation during the pilot study using various data recording methods, such as videos and photos. They defined an analytical framework comprising of seven design principles of mobilized curriculum and used the framework to analyse the observation data. They reported that there were significant changes in students' learning behaviours. Students engaged in more collaborative learning with peers and teachers, asked more questions, and became more creative.

Song et al. (2012) proposed a design process for creating a mobile-assisted personalized learning environment within Kolb's experiential learning model (Kolb 1984). The authors viewed that the mobility of learners enabled by mobile technologies would allow learners personalize both their learning goals (Jones and Issroff 2007) and context (Laurillard 2007). To overcome the limitations of experiential learning, Song et al. (2012) incorporated goal-based approach (Schank et al. 1994) into their mobile-assisted personalized learning environment to provide sustained experiential learning experience for science education. They then used design-based research methods to evaluate their design process on thirty-seven Primary 3 and 4 students in science inquiry. Data analysis conducted was based on artefacts produced by students during the evaluation process which involved formal and informal activities in class and out of class, such as activities performed in field trips. The authors concluded that their personalized learning goals and context enabled by the mobility of learners and re-designed curricular exhibited positive themes, particularly, themes related to the affordances of the mobile learning environment that they have created, such as enabling reflection and access to contents anytime and anywhere.

11.5 Types of Research Methodologies and Methods Utilised in m-Learning Research Papers

Table 11.1 summarizes learning types, for which m-learning researchers tried to integrate mobile technology. It shows that mobile technology can be applied to most of learning environments, such as in-class, out-of-class, and inquiry-based learning. It also showed that m-learning were applied to new emerging learning types as well, such as personalized learning, mobilized science curriculum, and social-interaction oriented learning.

Table 11.2 summaries research methodologies and evaluation approaches used by m-learning researchers. It shows that pilot studies utilizing observation, interviews, and questionnaires data gathering approaches were commonly employed to evaluate their proposed methods. Tables 11.3 and 11.4 summaries the data gathering approaches and data analysis approaches utilized in the evaluations.

Table 11.1 Learning types

Learning types supported	Papers
Project-based or inquiry-based learning	Zhang et al. (2010), Ting (2013), Song et al. (2012)
In-class learning	Zhang et al. (2010), Boticki et al. (2013)
Out-of-class learning	Song et al. (2012)
Personalized learning	Song et al. (2012)
Curriculum design: a methodology for designing mobilized science curriculum	Zhang et al. (2010)
Interwoven learning interactions	Ting (2013)
Social-interaction oriented learning	Ting (2013)
Formative assessment	Coulby et al. (2011), Hardyman et al. (2013)
Collaborative learning environment in classroom	Boticki et al. (2013)

Table 11.2 Research methodology and evaluation approaches

Evaluation approaches	Literature
Pilot-test curricular materials in classroom settings	Zhang et al. (2010), Boticki et al. (2013), Ting (2013)
Design-based	Song et al. (2012), Boticki et al. (2013)
Pilot formative assessment	Coulby et al. (2011)
Pilot training	Hardyman et al. (2013)
Ellaway's (Ellaway 2010) descriptive framework of e-learning	Hardyman et al. (2013)
David et al's (David et al. 2007) contextual mobile learning approach	Hardyman et al. (2013)
Student perceptions on m-learning	James (2011)

Table 11.3 Data Gathering Approaches

Data gathering approaches	Data collected	Literature
Observation	Artefacts	Zhang et al. (2010), Song et al. (2012)
	Note taking	Song et al. (2012), Zhang et al. (2010), Boticki et al. (2013)
	Video, image, notes	Zhang et al. (2010), Boticki et al. (2013)
In-direct observation	Narrative case reports	Hardyman et al. (2013)
	Data logging	Boticki et al. (2013)
Questionnaire, surveys	Questionnaire	Hardyman et al. (2013), Ting (2013)
Interview	Questionnaire, note taking	Coulby et al. (2011)
Focus-group study	Semi-structured questionnaire	James (2011), Boticki et al. (2013)
	Video	Boticki et al. (2013)
Summative evaluation on learning outcomes		Coulby et al. (2011), Zhang et al. (2010)

Table 11.4 Data analysis approaches

Data analysis approaches	Literature
Grounded theory approach using NVivo (coding)	Coulby et al. (2011), James (2011)
Case analysis	Zhang et al. (2010), Song et al. (2012), Coulby et al. (2011), Hardyman et al. (2013), Ting (2013)
Statistical analysis of questionnaire	Hardyman et al. (2013)
Statistical analysis of data logging or evaluations	Boticki et al. (2013), Zhang et al. (2010)

11.6 Conclusion: Gaps in Literature

In science learning, a typical lab setup creates an environment where students are required to solve a problem of replicating an experiment done by previous scientists, such as creating oxygen, setting up a computer network, and implementing a function for sending SMS messages. However, not all projects can have clear and direct outcomes to measure against the outcomes of the students. For example, designing a more fun and engaging interaction for an interactive product can be more challenging and time-consuming to assess. In both cases, however, the main challenges are stimulating students and facilitating effective learning within limited resources.

Previous studies suggest that m-learning creates learning environments where students are stimulated and effective learning is facilitated by providing more efficient personalized learning. However, these approaches mainly focus on re-designing curriculums, taking advantage of mobile technology, and improving communication and access to learning contents. Not much study has been done in collaborative experiential learning via mobile devices. Mobile collaborative experiential learning can allow students access a simulated lab anywhere at any time through short messages providing students a sense of ownership of their journey, experiential and contextualized learning in digital space, and formative assessment through continuous feedback from the simulated lab.

References

Boticki I, Wong LH, Looi CK (2013) Designing technology for content-independent collaborative mobile learning. IEEE Trans Learn Technol 6(1):14–24. doi:10.1109/tlt.2012.8

Churchill D, Churchill N (2008) Educational affordances of PDAs: a study of a teacher's exploration of this technology. Comput Educ 50(4):1439–1450

Conole G (2004) E-learning: the hype and the reality. J Interact Media Educ 2004(2):21. doi:10.5334/2004-12

Coulby C, Hennessey S, Davies N, Fuller R (2011) The use of mobile technology for work-based assessment: the student experience. Br J Educ Technol 42(2):251–265. doi:10.1111/j.1467-8535.2009.01022.x

David BT, Yin C, Chalon R (2007) Contextual mobile learning for appliance mastery. In: Proceedings of mobile learning:3–10

Devlin M (2012) Non-traditional student achievement: theory, policy and practice in Australian higher education. In: FYHE 2010 Aspiration, Access, Achievement: Proceedings of the 13th Pacific Rim First Year in Higher Education Conference 2010. Queensland University of Technology, Queensland, pp 1–13

Ellaway R (2010) Apples and architraves: a descriptive framework for e-learning research. Med Teach 32(1):95–97

Hardyman W, Bullock A, Brown A, Carter-Ingram S, Stacey M (2013) Mobile technology supporting trainee doctors' workplace learning and patient care: an evaluation. BMC medical education 13. doi:10.1186/1472-6920-13-6

James PT (2011) mobile-learning-thai HE student perceptions and potential technology impacts. Int Educ Stud 4(2):182

Jones A, Issroff K (2007) Motivation and mobile devices: exploring the role of appropriation and coping strategies. Res Learn Technol 15(3):247–258

Kneebone R, Brenton H (2005) Training perioperative specialist practitioners. Mobile learning: a handbook for educators and trainers. Routledge, Milton Park, pp 106–115

Kolb DA (1984) Experiential learning: experience as the source of learning and development, vol 1. Prentice-Hall Englewood, Cliffs

Kramer M (2009) Deciphering the future of learning through daily observation. Paper presented at the 3rd WLE mobile learning symposium: mobile learning cultures across education, work and leisure, WLE Centre, IOE, London, 27 March 2009

Laurillard D (2007) Pedagogical forms of mobile learning: framing research questions. WLE Center, London, pp 151–173

Lefoe GE, Hedberg J (2006) Blending on and off campus: a tale of two cities. In Bonk C, Graham CR (eds) The handbook of blended learning: global perspectives, local designs, pp 325–337

McInnis C, Hartley R (2002) Managing study and work: the impact of full-time study and paid work on the undergraduate experience in Australian Universities. DEST, Berlin

Norris C, Soloway E (2008) Getting mobile handhelds help bring K-12 classrooms into the 21st century. District Administration Magazine, Professional Media Group, LLC, CT. Available at http://www.districtadministration.com. Accessed 20 Oct 2010

Patten B, Arnedillo Sánchez I, Tangney B (2006) Designing collaborative, constructionist and contextual applications for handheld devices. Comput Educ 46(3):294–308

Schank RC, Fano A, Bell B, Jona M (1994) The design of goal-based scenarios. J Learn Sci 3 (4):305–345

Sharpe R, Benfield G, Roberts G, Francis R (2006) The undergraduate experience of blended e-learning: a review of UK literature and practice. Higher Education Academy, London

Song YJ, Wong LH, Looi CK (2012) Fostering personalized learning in science inquiry supported by mobile technologies. Educ Technol Res Dev (ETRD) 60(4):679–701. doi:10.1007/s11423-012-9245-6

Ting YL (2013) Using mobile technologies to create interwoven learning interactions: an intuitive design and its evaluation. Comput Educ 60(1):1–13. doi:10.1016/j.compedu.2012.07.004

Zhang B, Looi C-K, Seow P, Chia G, Wong L-H, Chen W, So H-J, Soloway E, Norris C (2010) Deconstructing and reconstructing: transforming primary science learning via a mobilized curriculum. Comput Educ 55(4):1504–1523. doi:10.1016/j.compedu.2010.06.016

Chapter 12
Important Factors in Choosing Tertiary Education Institutes

Abstract According to UNESCO's final figures for 2009, since 2000 the number of international students has increased by 75 %. The number of international students in Singapore is also expected to grow at 10 % annual rate reaching 150,000 in 2015. This number of students, according to a government's foresees, is worth USD 5 billion and contributes to 2 % of GDP. As a result, there are various great opportunities coming up for higher learning education in Singapore. Singapore has become one of the most favourite destinations for international students for higher education; its growth rate is surpassing the traditionally favoured Western countries, such as US. International students have also become an important economic factor. According to former researches related to education, students are usually influenced by the surrounding sources of information as well as their own experiences. Besides, institutional factors such as program type, reputation, location, and facilities are also considered when students choose their higher learning institutes. Other studies also recognized the extreme importance of several country factors such as living condition, and safety that may affect students when they consider a destination for their study abroad. This chapter aims to examine the sources of information that international students in Singapore consider in order to choose Singapore as their study abroad destination. Additionally, the study also compares the importance of country factors and institutional factors. We have surveyed 234 students who are currently studying in Singapore to find out the competitive advantage of Singapore Tertiary Education. The results show that recommendations from friends and teachers have major influence on students' decision in choosing their schools (higher education) and location of education, followed by country factors, such as personal security. Surprisingly, schools' promotional activities and agents have limited influence on students' decision making.

Keywords Asian century education · Singapore · Emerging market · International students

J. Vong and I. Song, *Emerging Technologies for Emerging Markets*,
Topics in Intelligent Engineering and Informatics 11,
DOI 10.1007/978-981-287-347-7_12

12.1 Introduction

In recent years, the number of international students around the world has increased dramatically in spite of global economic downturn. According to UNESCO's final figures for 2009, the number of international students rose by 75 % between 2000 and 2009. This figure reached 3.43 million in 2009, and is expected to increase annually by 12 % Coughlan (2011).

At present, Asia is the most vital source of international students due to economic boom in several Asian areas, such as China, India and Southeast Asia. According to the UK Vision 2020 report, Asia is set to account for 70 % of the global demand for international higher education by 2025.

While Western countries, such as USA, UK, Australia, still remain as the favourite destinations, with great demand and development of several Asian countries in education, a new trend is starting to emerge. Based on the Times Higher Education World Reputation Rankings (Morgan 2011), many of Asia's institutions have become mainstream top universities in the world, making Singapore, Japan, Korea, and even China as potential destinations for study abroad. In particular, Singapore places strategic intent to increase international students, such as the 'Global Schoolhouse' blueprint (Synergy 2011). Singapore is now expecting to host 150,000 foreign students by 2015, creating 22,000 jobs and contributing approximately USD 5 billion to the economy, about 2–5 % of Singapore GDP (TheBusinessTimes 2008). Therefore, there are great opportunities to invest in education industry. In a business context, understanding customer behaviour is very important for firms to design their own business strategies. According to Philip Kotler (2012), in order to create value for customers and build meaningful relationship with customers, marketers must first gain fresh and deep insight into what customers need, and this is also critical in developing competitive advantages for the firm.

The main objective of this case study is to discover the factors that international students consider in choosing their schools. Unlike previous studies, data is collected from students who are currently studying in many different public and private universities in Singapore. The case study focuses on proving the following two key hypotheses that are important for the market research of educational sectors:

- Hypothesis 1: Surrounding relationships affect more on students than students' personal experiences when students consider their study abroad.
- Hypothesis 2: Institutional factors are more important reasons for students than country factors when deciding to study in Singapore.

The paper is organized as follows. In the next section, we review previous studies in educational service markets. In Sect. 12.3, the research methodology is de-scribed. In Sect. 12.4, we summarize the survey data collected for the study. In Sect. 12.5, the results of the analysis are shown. Finally in Sect. 12.6, discussion and conclusion are made.

12.2 Background

12.2.1 Types of Information Sources Affecting International Students

According to Philip Kotler (2012), international students rely on the following types of information sources in choosing educational institutes: personal sources (family, friends, acquaintances, etc.), commercial sources (advertising, sales people, etc.), public sources (Internet search, firm ranking, etc.) and experiential sources. An early study done by Steadman and Dagwell (1990) showed that the opinions of family and friends have major influence on students' choice of colleges. In a paper by Donnellan (2002) that explores the impact of marketer controlled factors on college-choice decisions by students at a public research university, it was concluded that, beyond the influence of parents, other factors such as personal contacts and social life have great impact on students' decisions. Similarly, Raposo (2007) concluded that personal perception, including background, financial ability and culture, have the greatest impact while the influence of external factors, such as advertisements, has negligible impact on the decision-making. Besides, parents and school teachers' recommendations strongly influence in the process of selecting a university. However, students depend more on a variety of information sources for their decision-making, particularly, information available on the internet (Tricker 2003). This factor has to be taken into account while examining the driving factors of students' choice of higher education program and destination.

12.2.2 Institutional Factors in Choosing Study Abroad Destinations

Many researches and propositions related to students' choice of higher learning institutes indicate that factors related to universities or higher learning institutes play a very important role in student decision-making process. Chapman (1981) proposed a model of students' decision making process in selecting colleges. The model emphasizes three main factors: (1) quality of faculty; (2) quality of degrees; and (3) overalls academic reputation. These factors are mainly based on educational quality of institutes. Coccari and Javalgi (1995) pointed out that, in addition to the educational quality factors, cost and variety of courses offered have an impact on student's decision. In addition, a research paper by Kallio (1995), it was highlighted that institutional quality, reputation and course diversity as factors that affect the graduate students' choice of college. Apart from these factors, it also reported the importance of external and environmental factors such as residency, academic environment and financial aid. Another research conducted by Soutar GaT (2002) mentioned that students are now more concerned about the bottom line outcome after they complete their degrees. This suggests job prospect and demand of relevant skills in workplace are also some of the factors affecting their choice.

12.2.3 Country Factors in Choosing Study Abroad Destinations

Nowadays, students have a broad range of alternatives for their higher learning education. The choice is not limited within domestic institutions; however, it has been spread to global platform.

Shanka and Quintal (2005) examined the major reasons why students chose Australia. They reported that the distance of the school to a student's home country is a vital factor that influences international students' choice followed by safety, educational quality, cost of living, and tuition fee. Similarly, Singh (2009), the executive director of Angad Info—Overseas Communication Consultancy, stated that safety and cultural similarities are Singapore's advantages that ensures in bringing international students, especially from India and regional areas. Yilin (2010)'s study on Chinese students' decision making of overseas studies claimed that those who are willing to study abroad have strong motivation to live and work overseas after graduating. This indicates that students are more independent in choosing study abroad destinations. Cost of study abroad is a big issue for students. In a study by Anderson and Bhati (2012), it was revealed that safety, location of study destination, lower cost of living are some of the important factors for Indian students in determining their choice. The perception of better job prospects also affects their choice.

Overall, institutional factors, such as reputation, academic quality, facilities, tuition fee, and country factors, such as safety, living cost, career, and settlement opportunities, have an influence on student's decision. Hence, this paper aims to focus on examining separately the international students' perception of country factors and institutional factors so as to come up with a specific set of reasons for students to choose Singapore as their study abroad destination. Besides that, the sources of information that students use to consider their studies abroad have also been explored.

12.3 Methodology

We had surveyed international students who are currently studying in Singapore. A questionnaire was designed to gather the following types of information from the international students using a five-point Likert scale. There were three parts in the questionnaire.

The first part of the questionnaire asked for some personal information, such as country of origin of the respondents, university or college they are attending, and course they are doing (diploma, foundation, bachelor, Master or PhD);

The second part of the questionnaire contained a ranking question where the respondents had to rank the top 10 countries where they had ever considered for their study abroad.

The third part of the questionnaire asked respondents to rate from the scale of 1 (strongly disagree) to 5 (strongly agree) according to the importance of factors that influence their decision in choosing Singapore as their study abroad destination. There were three areas:

1. Information sources such as agent and the internet.
2. Country factors, such as living condition and safety.
3. Institutional factors, such as campus facilities, institute reputation, and student services.

The final part of the questionnaire asked respondents to rate on the overall satisfaction level of students in Singapore on the scale from 1 (strongly disagree) to 5 (strongly agree). Data collection was conducted from the 8th of August to 17th of August in 2012. The respondents were diploma, foundation, bachelor, master and PhD students and they were randomly chosen from various different public and private universities across Singapore which includes SIM, JCU, NUS, NTU and other private educational institutes. The questionnaires were completed in two ways: online and by delivery. For online survey, the questionnaire was posted in Survey-Monkey, and the link of the questionnaire is sent to the international students in Singapore. The questionnaires were also distributed in the universities that were targeted.

Quantitative analysis had been used to analyse and interpret the case study result: The first, second and fourth parts of questionnaire that were related to demography information, students' preferences of study abroad destination, and their overall satisfaction of studying in Singapore were represented through graph forms to give an entire view of respondents.

The third part data was analysed by using quantitative statistic methods. Since the importance of each given factor was rated by respondents on the scale of 1 (strongly disagree) to 5 (strongly agree), the paper's method is to compare mean of each factor with 3 (neutral) by using t-test which is one of the statistical method that determines how significant is the difference. In this case, the hypothetical mean is 3. So if the mean of observed factor is statistically higher than 3, that factor is considered as having significant effect on students' decision. Furthermore, two sample t-test were used to compare the significance of difference between rated factors. This is also a statistical method which calculates a value that determines the significance of the difference in results. The interpretation of two sample t-test resulted in a priority order of factors. The test was conducted in terms of entire sample and home countries to unveil the differences in consideration of students from different parts of the world. Finally, two sample t-test was used to find out whether country factors are relatively more significant than institutional factors or not. The analysis was conducted separately with different group of students who are from different parts of the world in order to identify the distinctive features of each group of students. All t-tests have been conducted with 5 % of significant level.

However, there were several factors that may lead to the inaccuracy of the case study results. For example, the respondent sample may not represent entire population of international students in Singapore. In addition, the given factors in each theme may not represent all the possible options. Furthermore, some misunderstanding of questionnaire content may also take place.

12.4 Survey Analyses

12.4.1 Demographic Information of Respondents

Respondents of case study are from various higher learning institutes in Singapore.

Table 12.1 shows the demographics of the respondents. A total of 234 survey responses are collected. About 24 % of respondents are from public universities and the majority of the rest are from private higher learning institutes. Also, the majority of the respondents are from South East Asia and China which account for more than 60 % of total numbers of respondents. Few respondents are from countries like Europe, USA, Australia, Japan, and South Korea too.

It can be seen from Fig. 12.1 that majority of respondents are from Bachelor's courses, and this is followed by master's degree. On the other hand, the minority are from diploma, PhD and foundation courses.

Table 12.1 Demographics of respondents

	India	China	SE Asia	Other	Sum	%
NUS	7	31	4	2	44	19
NTU	3	2	3	0	8	3
JCU	22	21	39	26	108	46
SIM	0	5	6	4	15	6
MDIS	0	0	2	3	5	2
PSB	4	7	4	7	22	9
NYP	0	3	5	0	8	3
Other	3	4	11	6	24	10
Sum	39	73	74	48	234	100
%	17	31	32	20	100	

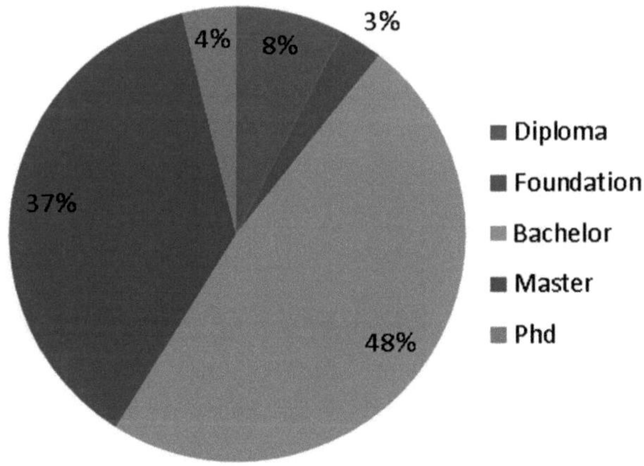

Fig. 12.1 Respondents according to course

12.4.2 References of Study Abroad Destination

The survey result in Table 12.2 shows that top 5 countries which are usually considered for overseas education are USA, UK, Australia, Canada, and Singapore (in order of top 1 top 5). The outcome is interesting. Singapore was the only Asian country in the top 5 ranking.

12.4.3 Importance of Sources of Information for Prospective Students

As shown in Fig. 12.2, we defined seven sources of information and classified into 2 groups:

1. Surrounding relationship: Suggestion from parent; former teachers; friends.
2. Personal experience: Information from Internet, promotion activities of higher learning institutes such as Event and Advertisement, Suggestion of Agent.

Table 12.2 Countries considered for overseas education

	Priority of choices					
	1st	2nd	3rd	4th	5th	Total choices
USA	(1)	(2)	(3)	(4)	(5)	(6) = (1) + (2) + (3) + (4) + (5)
	75	53	29	12	2	171
UK	51	53	39	14	5	162
Australia	47	41	42	13	3	146
Canada	14	11	13	15	7	60
Singapore	**11**	**12**	**12**	**12**	**11**	**58**

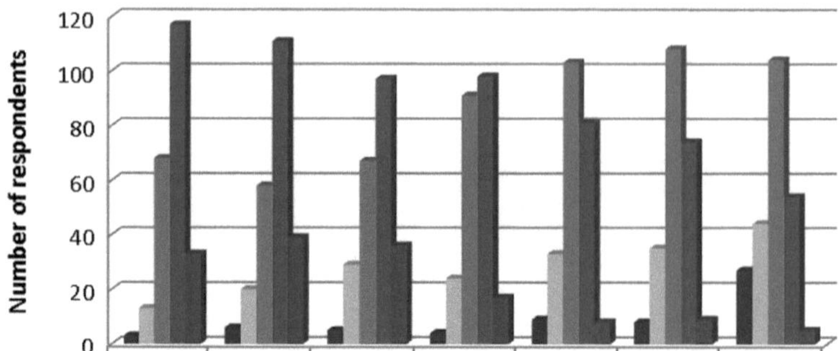

	Internet	Parents	Teacher	Friends	Event	Advertisement	Agent
Strongly Disagree	3	6	5	4	9	8	27
Disagree	13	20	29	24	33	35	44
Neutral	68	58	67	91	103	108	104
Agree	117	111	97	98	81	74	54
Strongly Agree	33	39	36	17	8	9	5

Fig. 12.2 Importance of sources of information

Figure 12.2 shows that the majority of the respondents (more than 50 %) rated information sources from internet, parents, and teachers as "agree" or "strongly agree". Meanwhile, most of them (45 %) are neutral, disagree or strongly disagree with the importance of event, advertisement and suggestion of Agent by rating these factors as 'Neutral'. To prove the correctness of this interpretation, a statistical method is used known as one sample t-test (significant level is 5 %) which also gave the similar result as shown in Table 12.3.

Table 12.4 shows that the results seem to be similar when one sample t-test is repeated with sub-sample of respondents coming from different part of the world,

Table 12.3 Sources of information one—sample test with test value = 3

	t	df	Sig. (2-tailed)	Mean difference	95 % Confidence interval of the difference	
					Lower	Upper
Internet	12.976	233	0.000	0.701	0.59	0.81
Parents	10.924	233	0.000	0.671	0.55	0.79
Teacher	8.790	233	0.000	0.556	0.43	0.68
Friends	7.810	233	0.000	0.427	0.32	0.54
Event	3.490	233	0.001	0.197	0.09	0.31
Advertisement	3.140	233	0.002	0.175	0.07	0.29
Agent	−2.282	233	0.023	−0.145	−0.27	−0.02

Table 12.4 One-sample test—agent with test value = 3

	Mean difference	95 % Confidence interval of the different	
		Lower	Upper
China	−1.64	−0.40	0.07
Southeast Asia	−0.041	−0.26	0.17
India	−0.205	−0.53	0.12
Others	−0.229	−0.51	0.05

such as China, Southeast Asia, India and others. "Agent" was the only factor that can insignificantly influence students' decision. This significance was tested for each country for the factor "agent", as follows.

Using two sample t-test to recognize the difference between each pair of factors, the result shows that, in most of the cases, the difference between internet and other factors related to students' relationship is insignificant. However, the significant difference comes from two groups of factors: "Internet, Parents, Teachers, Friends" and "Advertisement, Event, Agent". It also released the same result when testing is repeated with personal country.

12.4.4 Importance of Country Factors

The country factors considered for the survey were: living condition, living cost, safety, job and settlement, education quality guarantee, location, cultural similarity and international experiences.

According to the Fig. 12.3, more than 60 % of students agree and strongly agree that country factors are important for considering study abroad destination. Out of 234 questionnaires collected, 209 students (89 %) agree and strongly agree that safety is the most important factor for them followed by living conditions, education quality guarantee and job and settlement with each obtained 72, 68 and 66 % of the total students respectively in "agree" and "strongly agree". Only 53 % of the students voted cultural similarity thus it was the weakest influential factor.

However, to ensure the correctness of result, it is necessary to use a statistical method known as one sample t-test and the result revealed is as shown in Table 12.5.

The one sample t-test demonstrated that all factors mentioned above influence the students' decision to receive education in Singapore. Interestingly, among the 8 factors listed, safety is the most important factor for the students. According to the test, it could be divided into 3 main scales according to the level of significant impact in general:

- Strong impact: safety
- Intermediate impact: living condition, education quality guarantee, job and settlement, international experience and location
- Slight impact: cultural similarity and living cost

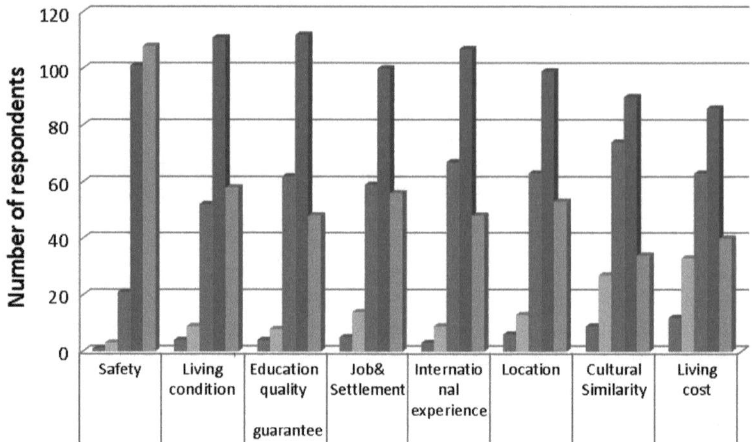

	Safety	Living condition	Education quality guarantee	Job& Settlement	International experience	Location	Cultural Similarity	Living cost
Strongly Disagree	1	4	4	5	3	6	9	12
Disagree	3	9	8	14	9	13	27	33
Neutral	21	52	62	59	67	63	74	63
Agree	101	111	112	100	107	99	90	86
Strongly Agree	108	58	48	56	48	53	34	40

Fig. 12.3 Importance of country factors

Table 12.5 Country factors one-sample test with test value = 3

	t	df	Sig. (2-tailed)	Mean difference	95 % Confidence interval of the difference	
					Lower	Upper
Safety	27.958	233	0.000	1.333	1.24	1.43
Living condition	15.639	233	0.000	0.897	0.78	1.01
Education quality guarantee	14.674	233	0.000	0.821	0.71	0.93
Job and settlement	13.040	233	0.000	0.803	0.68	0.92
International experience	14.432	233	0.000	0.803	0.69	0.91
Location	12.408	233	0.000	0.769	0.65	0.89
Cultural similarity	7.364	233	0.000	0.483	0.35	0.61
Living cost	6.544	233	0.000	0.466	0.33	0.61

To elaborate, Chinese students highlight job and settlement opportunities as the second most important reason in making their decision whereas students from Southeast Asia prefer living condition and education quality guarantee as their reasons. Besides, students from other countries (Europe, USA, Australia, Korea, Japan, etc.) selected living condition, international experiences and convenience in location as their most considered factors regarding Singapore education.

12.4.5 Importance of Institutional Factors

Figure 12.4 represents the distribution of respondents' opinion in term of importance level of institutional factors, it shows that 61 % of students agree and strongly agree that these factors are important for them to choose their study abroad destination. Interestingly, the institute's program is considered as the most important factor for students in making their decision. It is accounted for 74 % of the students surveyed (total 174 out of 234 students voted "agree" or "strongly agreed"). The second most important factor is the reputation of institutes in which 73 % of the students agreed or strongly disagreed and this is followed by the factor, facilities, with 150 out of 234 students voted (64 %). On the other hand, students seem to least consider campus activities. Only 42 % of students agree and strongly agree that it is important factor and 47 % of the students (109 students) voted "neutral". These factors are tested by using one sample t-test as shown in Table 12.6, to ensure the correctness of results and their interpretation:

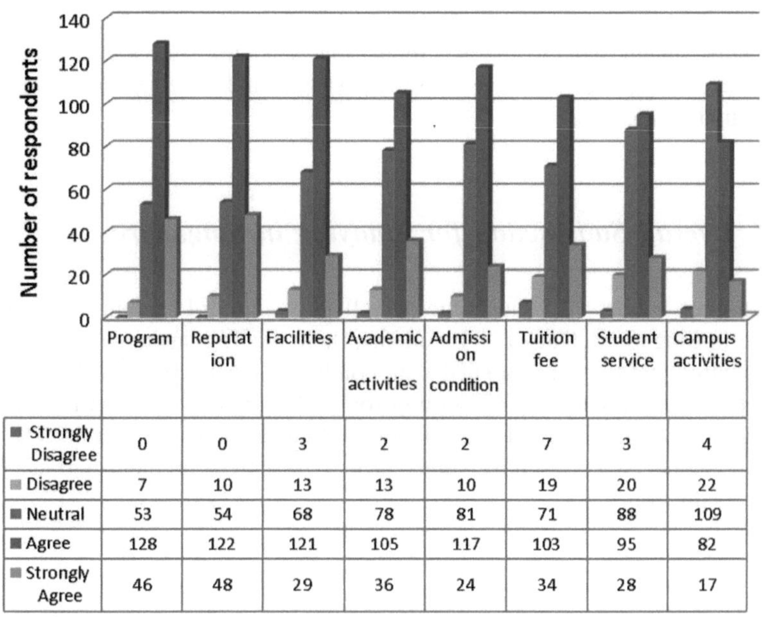

	Program	Reputat ion	Facilities	Avademic activities	Admissi on condition	Tuition fee	Student service	Campus activities
Strongly Disagree	0	0	3	2	2	7	3	4
Disagree	7	10	13	13	10	19	20	22
Neutral	53	54	68	78	81	71	88	109
Agree	128	122	121	105	117	103	95	82
Strongly Agree	46	48	29	36	24	34	28	17

Fig. 12.4 Importance of institutional factors

Table 12.6 Institutional factors one-sample test with test value = 3

	t	df	Sig. (2-tailed)	Mean difference	95 % Confidence interval of the difference	
					Lower	Upper
Program	19.002	233	0.000	0.910	0.82	1.00
Reputation	17.598	233	0.000	0.889	0.79	0.99
Facilities	12.925	233	0.000	0.684	0.58	0.79
Academic activities	12.599	233	0.000	0.684	0.58	0.79
Admission condition	13.038	233	0.000	0.645	0.55	0.74
Tuition fee	9.627	233	0.000	0.590	0.47	0.71
Student services	9.505	233	0.000	0.534	0.42	0.64
Campus activities	6.862	233	0.000	0.368	0.26	0.47

As oppose to the test result, all factors including Campus Facilities, Institutional reputation, program content, academic activities, campus activities, tuition fees, admission condition, and student service have an influence on students' decision. According to the result, program contents and institutional reputation are generally the most important factors while other factors such as facilities, academic activities, admission condition, tuition fees are also highly important but the last two factors student services and campus activities are less important.

For Indian students, academic activities and institutional reputation are the two most important concerns followed by program content. For Chinese students, next to program content and institutional reputation, academic activities are important factor whereas students coming from Southeast Asia and Others are more interested in Campus facilities.

12.4.6 Overall Satisfaction for Studying in Singapore

The histogram in Fig. 12.5 shows the overall satisfaction of students on living and studying in Singapore. Out of 234 students, there are 154 students who agree and strongly agree that they are satisfied with condition in Singapore, making 66 % of the sample size. 69 students, accounting for 29 %, are less satisfied in living and studying in Singapore. Also, 11 respondents which accounts for around 5 % of the total respondents chose "disagree" and "strongly disagree" on their grading of their overall satisfaction.

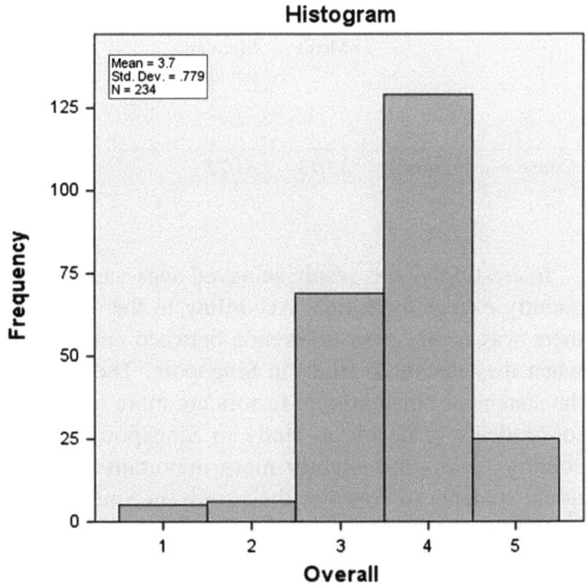

Fig. 12.5 Overall satisfaction in studying in Singapore

12.5 Hypothesis Testing

Hypothesis 1: "**Surrounding relationships affects more on students than students' personal experiences when students consider their study abroad**."
The case study unveiled that information from internet and suggestions of parents, former teachers and friends mostly impact on students' decision making process. In most of the cases, the differences within these factors are not significant. On one hand, to gather information from Internet is a personal research process that partly reflects students' personal experience. On the other hand, suggestion of parents, former teachers and friend reflect the impacts of surrounding relationship.

Therefore, the case study result allows rejection of the statement "surrounding relationships are more important sources of information than students' personal experiences when students consider their study abroad". In fact, both surrounding relationship and personal experiences has high influence on student's decision making.

Hypothesis 2: "**Institutional factors are more important reasons than country factors for students to decide on studying in Singapore**."
In general, both country factors and institutional factors are significantly important. Two sample t-test (Paired sample test) was used to check if there is any significant difference between country factors and institutional factors. In the test, the sum of grade of eight country factors was compared to the sum of grade of eight institutional factors as shown in Table 12.7. The result shows that country factors have slightly greater impact than institutional factors.

Table 12.7 Country factor—institutional factors paired-sample test (paired differences)

	Mean	Standard deviation	Standard error mean	95 % Confidence interval of the difference	
				Lower	Upper
Country-institutional	1.073	4.027	0.263	0.554	1.591

Interestingly, the result achieved was same when repeating the test for each country except for China. According to the data collected from Chinese students, there was nearly zero difference between country factors and institutional factors when they decide to study in Singapore. The case study results allow rejection of the statement "Institutional factors are more important reasons than country factors for students to decide to study in Singapore". In fact, in the case of Singapore, country factors are slightly more important than institutional factors for international students to consider their study in Singapore.

12.6 Discussion

The case study information can be very useful in designing communication, academic and campus strategies for higher learning institutes in Singapore in order to attract more international students and also to come up to their expectations.

In terms of communication strategies, according to the case study result, the internet is the most important source of information; therefore, higher learning institutes need to concentrate on this information channel. The information of academic program and their application into the real workplace must be highlighted at institute's website, since this is the basic information, according to the research result. Furthermore, the name of institute needs to appear in other well–known academic and informative websites as well as in ranking list of the particular program. The greater reputation an institute has, the more attraction could be created. In fact, to achieve those 'public relation' effects, institutes have to devote many efforts in research and getting involved in local as well as global economic and academic life.

Last but not the least, to have a close relationship with alumni can enhance the positive word about institutes. This also plays an important role in building institute's reputation since students tend to be strongly influenced by their personal relationship's suggestions. In terms of implication for campus activities, the study results suggest in designing attractive academic and campus activities.

Students usually expect to experience the real 'Singapore life' during their study abroad journey. Hence, the academic curriculum of higher learning institutes in Singapore should reconcile to the current local economic and social life.

Furthermore, institutes need to encourage multi-national teamwork and environment whereby students coming from different parts of the world are able to learn from diversity of each other, so as to enhance cultural exchange and learning process.

12.7 Conclusion

Singapore is coming up as one of the most favourite destinations for higher studies and majority of students studying here are satisfied with their decisions. However, the country is being challenged by other destinations such as USA, UK, Australia, Canada as well as emerging destinations such as Japan, Korea and Hong Kong.

The internet and personal relationships are the most important sources of information for students to consider their study destination. With respect to country factors, safety is most important while living condition, education quality, job& settlement opportunity, international experiences and location are also vital reasons for student to choose Singapore as their study destination. As far as institutional factors are concerned, program offered and institute reputation are most important factors for most of the students followed by campus facilities, academic activities, admission condition, and tuition fee.

Both country factors and institutional factors play a vital role for students to decide their study destination, but in case of Singapore, country factors tend to be slightly more important than institutional factors. Based on results of survey, higher learning institutes and government in Singapore should concentrate more on using Internet and alumni community for communication activities, and country's specific features as well as education quality and institutes' reputation should also be highlighted well. Furthermore, higher learning institutes in Singapore should provide a multi-national environment, and should design their academic and campus activities that aligns to Singapore's experience.

For further research, studies should concentrate on comparing competitive advantages of Singapore higher learning institutes with competitors like Northern America, Europe and Australia.

References

Anderson R, Bhati A (2012) Indian students' choice of study destination: reasons for choose Singapore. Int J Innovative Interdisc Res 1(2):66–76

Chapman DW (1981) A model of student college choice. J High Educ 52(5):490–505

Coccari R, Javalgi R (1995) Analysis of students' needs in selecting a college or university in a changing environment. J Mark High Educ 6(2):27–40

Coughlan S (2011) Records number of international students. BBC New Business. http://www.bbc.co.uk/news/business-12671198

Donnellan J (2002) The impact of marketer controlled factors on college-choice decisions by students at a public research university. Graduate School of the University of Massachusetts Amherst
Kallio RE (1995) Factors influencing the college choice decisions of graduate students. Res High Educ 36(1):109–124
Morgan J (2011) Times higher education world reputation rankings. Times Higher Education. http://www.timeshighereducation.co.uk/story.asp?storycode=415429
Philip Kotler GA (2012) Principles of Marketing. Pearson, New Jersey
Raposo M (2007) A model of university choice: an exploratory approach. MPRA. http://mpra.ub.uni-muenchen.de/5523/
Shanka T, Quintal V (2005) Factors influencing international students' choice of an education destination—a correspondence analysis. J Mark High Educ 15:31–46
Singh R (2009) Singapore is a safe study destination. Get ahead http://getahead.rediff.com/report/2009/jun/23/singapore-is-a-safe-study-destination.htm
Soutar GaT JP (2002) Students' preferences for university: a conjoint analysis. The Int J Educ Manag 16(1):40–45
Steadman GT, Dagwell RH (1990) A survey of overseas students in Australia. Aust Univ Rev 33 (1):59–63
Synergy (2011) Singapore—the Global Schoolhouse. Synergy SEP/OCTCT 2011
TheBusinessTimes (2008) Foreign units bet big on a Singapore School. The Business Times
Tricker T (2003) Student Expectations—How do we measure up? Students Living Report. School of computing and management sciences, Sheffield Hallam University, UK
Yilin L (2010) Factors influencing the choice of overseas study by undergraduate and postgraduate Chinese students. www.duplication.net.au/ANZMAC09/papers/ANZMAC2009-740.pdf

Chapter 13
Automated Tutoring System

Abstract This chapter reports a new simulated smart learning environment, called Mobile Collaborative Experiential Learning (MCEL). MCEL provides automated, continuous, personalized, and formative feedback. Students interact with the simulated learning system via simple text messages using mobile devices in order to change the state of the system to a desired state over time. In the process, students engage in complex problem solving activities, and the system provides continuous formative assessment to help the students achieve learning objectives. Unlike conventional summative assessment approaches, students acquire set competency levels during their journey to achieving certain goals while interacting with the simulated system. MCEL allows instructors easily define learning journeys using event-condition-action (ECA) rules and reliable structured-text parser. A pilot study of MCEL is conducted and evaluated on a group of twenty Masters-in-IT students. Their participation is observed and a survey is conducted. The evaluation results show that MCEL supports Bigg's constructive alignment in curriculum design, contextualized experimental learning, and personalized formative learning.

Keywords Mobile education · M-learning · SOA 2.0 · ECA · Automated tutoring

13.1 Introduction

In science learning, a typical laboratory setup would expect students to replicate the results of certain experiments done by previous scientists, such as creating oxygen, setting up computer networks, and implementing software functions for sending SMS messages. However, not all student projects can have clear and direct outcomes to measure against the outcomes of students. For example, designing a more fun and engaging interaction for an interactive product can be more challenging and time consuming to assess. In both cases, however, the main challenges are stimulating students, facilitating effective learning, extend learning beyond classrooms within limited resources in the age of massification of education market.

J. Vong and I. Song, *Emerging Technologies for Emerging Markets*,
Topics in Intelligent Engineering and Informatics 11,
DOI 10.1007/978-981-287-347-7_13

Researchers have been trying solve these challenges through flexible learning (McInnis and Hartley 2002; Devlin 2012; Lefoe and Hedberg 2006) and blended learning (Sharpe et al. 2006) approaches. Faced with new digital age generations (Conole 2004), unprecedented network speed and mobile technology, education researchers now frequently quote mobilized curriculum (Zhang et al. 2010), social-interaction oriented learning (Ting 2013), personalized learning (Song et al. 2012), and collaborative mobile learning (Boticki et al. 2013) in their studies. These previous studies suggest that mobile learning approaches can be used to create learning environments that can stimulate students, facilitate effective learning, and extend learning beyond classrooms by providing personalized learning more efficiently.

However, these studies have focused mainly on re-designing curriculums taking advantage of mobile technology, and improving communication and access to learning contents. In this paper, we focus on exploring how mobile devices can be used to create collaborative experiential learning environment. We envision that mobile learning can allow students to access simulated learning laboratories anywhere any-time through mobile devices, providing experiential and contextualized learning in digital space, and formative assessment through continuous feedback from the simulated lab. The objectives of this new research direction are to (a) improve student engagement; (b) provide ownership of their learning journey; and (c) provide experiential, contextualized and formative assessment at low cost.

To achieve these goals, Mobile Collaborative Experiential Learning (MCEL) has been developed using design-based research approach (Brown 1992). MCEL combines service-orientation design paradigm and context-free grammar parser to provide easy to use curriculum development platform and natural-language-based user interaction. Teachers can express learning contents and objectives using event-condition-action rules and students can interact with MCEL using structured SMS message. A pilot study was conducted to evaluate the effectiveness of MCEL. The evaluation results show that MCEL is usable, students' learning activity continued beyond classes, students received feedback anywhere anytime, and they felt sense of ownership of their learning journey.

The chapter is organized as follows. In Sect. 13.2, we overview the core techniques for providing automated tutoring services. In Sect. 13.3, we describe development of MCEL, pilot-study, and evaluation approaches designed based on the literature review. In Sect. 13.4, we report the evaluation results. In Sect. 13.5, we conclude the paper with remarks.

13.2 Background

This work is based on m-learning, collaborative learning, social interaction in learning, mobile technologies, service-orientation design paradigm, and natural language parsing. The following sections provide a brief overview of the core techniques, focusing on mobile learning for mobilized curriculum, formative assessment, and enhancing social experiences.

13.2.1 Mobile Learning for Enhancing Social Experiences

One primary goal of educators is creating a harmonized environment where students could construct their learning upon both physical and social contexts augmented on the learning content. Therefore, social interaction is considered as the key advantage and ingredient in the designs of mobile learning practices. In particular, some researchers viewed that mobile learning could enable students to interact with peers, teachers, and subject contents more efficiently and unobtrusively (Ting 2013). Furthermore, Boticki et al. (2013) proposed that mobile-learning can provide efficient collaborative learning environment to enhance in-class learning experiences.

Based on these views, researchers proposed a mobile-technology platform for supporting collaborative learning in class and evaluated the platform in teaching two Primary 3 subjects in class: Chinese characters and mathematics (Boticki et al. 2013). On the other hand, Ting (2013) proposed a curriculum design practice, called interwoven learning interactions, for creating more socially interactive learning environments.

13.2.2 Technologies for Automating Assessment

Now, over 87 % of the global population and 79 % of developing world's population has access to Simple Messaging Service (SMS) (ITU 2011). This pervasiveness of mobile technologies has created both challenges Kneebone and Brenton (2005) and opportunities (Coulby et al. 2011). As seen in previous sections, we are still learning how to integrate these new technologies. In particular, we need to develop platforms for designing curriculums, methodologies for re-designing existing curriculums (Zhang et al. 2010), and automated methods for providing formative assessments (Coulby et al. 2011). One particular technology that has been proven to be intuitive and efficient in capturing and implementing business requirements in business process automation is event-condition-action (ECA) rules. ECA is the basis of event-driven architecture (EDA) pattern adopted by SOA (Service-Oriented Architecture) 2.0 for more natural modelling of business processes. ECA can be converted into Business Process Execution Language for Web Services (BPEL) (Ouyang et al. 2006), allowing an easy migration between different modelling tools. BPEL is supported by major business process management software and development tools (e.g., Eclipse IDE).

Formal language processing (Roche and Schabes 1997) has been proven efficient and effective in processing structured languages, such as formulas and statements of programming languages. Its advantage over natural language processing (Manning and Schütze 1999) is their reliability and accuracy, which are very important usability aspects. Re-searchers can now build custom grammars using lexers and parsers [e.g., ANTLR (Parr 2007)] defined using simple rewrite rules to automatically process structured SMS messages.

13.3 Mobile Collaborative Experiential Learning (MCEL) System

These previous studies suggest that m-learning can be used to create learning environments that can stimulate students and facilitate effective learning by providing personalized learning more efficiently. However, these approaches mainly focused on re-designing curriculums by taking advantage of mobile technology, and improving communication and access to learning contents.

13.3.1 Methodology

MCEL is developed and evaluated using design-based research approach (Brown 1992). The initial prototyping is done using wizard-of-oz prototyping, in which human instructors simulate the automated system using structured instructions and rules. After the usability and technical feasibility problems are addressed, the system is implemented on an Android tablet. Figure 13.1 shows the overall system diagram of MCEL, which is a simulated lab running on an Android tablet. It can continuously interact with students in order to deliver personalized assessment via simple text messages. The learning script is formulated based on Event-Condition-Action (ECA) paradigm. An Android tablet with Android OS v3.0 is used to host the simulated lab. ANTLR (Parr 2007) is used to generate simple lexer-parser Java-code for our structured SMS messages.

For the evaluation, a learning script for two questions was prepared. Students are told that an automated system will guide their learning by interacting with them through simple text messaging services. Survey questions are designed based on previous mobile education studies done by Nokia in 2008. Three sets of questions are selected from the question set used by these previous researchers as shown in Table 13.1: (1) User-friendliness, (2) Technical feasibility, and (3) Cost effectiveness.

Survey results are plotted on a stacked bar chart in Fig. 13.2. This chart shows that most of the participants viewed MCEL useful (user-friendly both in learning and technical accessibility) and cost effective.

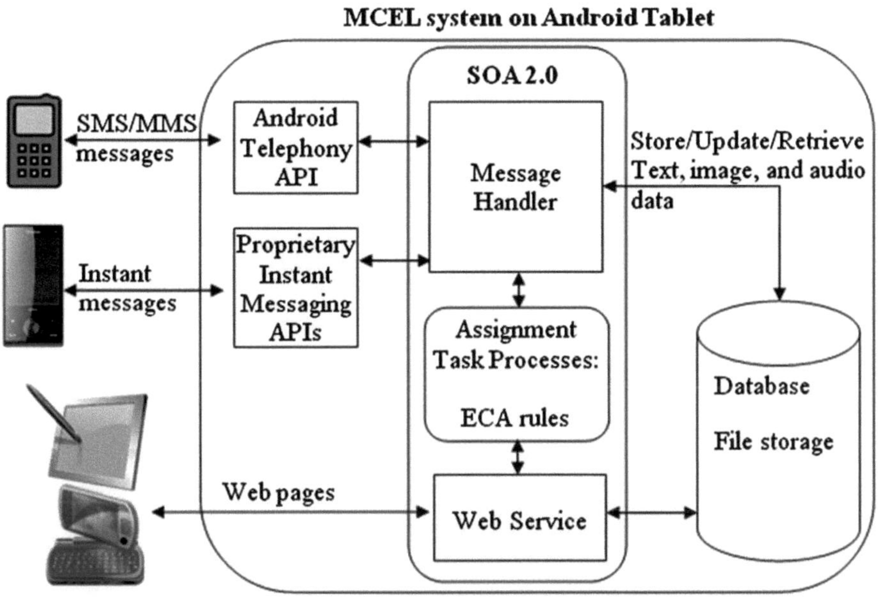

Fig. 13.1 Mobile Collaborative Experiential Learning (MCEL) System

Table 13.1 Survey questions designed on the perceived usefulness, usability, technical feasibility, and cost effectiveness of students

Question no.	Questions	Avg.
Student user-friendliness (5-point Likert scale response: (1–Strongly disagree, 2–agree, 3–Neutral, 4–agree, 5–Strongly agree)		
U1	It was easy to use the equipment in this mobile learning course	4.14
U2	This mobile learning experience was fun	4.29
U3	According to my experience I would take another mobile learning course if relevant to my learning needs	4.43
U4	I would recommend mobile learning as a method of study to others	4.29
U5	Mobile learning increases the quality of e-learning	3.86
U6	Course learning objectives can be met by mobile learning	3.43
U7	Communication with and feedback from the tutor in this course was easy	4.29
U8	Mobile learning may be convenient for communication with other course students	4.14
Technical feasibility (5-point Likert scale response)		
T1	Navigation through the mobile learning course was easy	4.14
T2	For mobile learning to be effective it is necessary to use graphics and illustrations	4.57
T3	Evaluation and questioning in the mobile learning course was effective	4.14
Cost effectiveness (5-point Likert scale response)		
C1	Mobile learning increases access to education and training	4.00
C2	The cost of communicating in the mobile learning course with the tutor and other students was acceptable	4.00

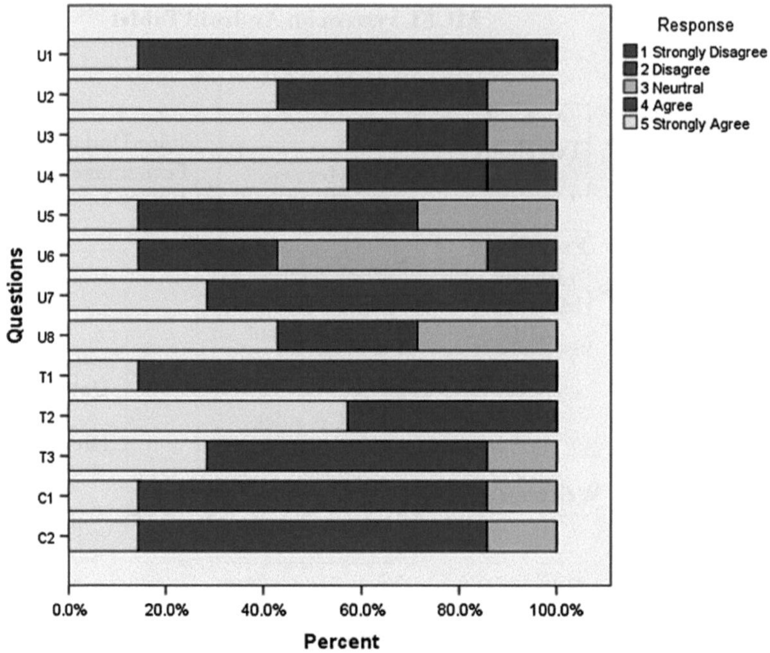

Fig. 13.2 Survey results plotted on a stacked bar chart. This chart shows that most of the participants viewed MCEL useful (user-friendly both in learning and technical accessibility) and cost effective

13.4 Experimental Results

Figure 13.3 shows sample text messages being exchanged between MCEL and students. Some students sent replies even at midnights, clearly showing evidence of their interest and engagement. Out of 22 participants, 6 students completed the questionnaires. Table 13.1 and Fig. 13.2 shows the survey results, which demonstrate that most of the participants viewed MCEL useful: user-friendly in terms of both learning and technical accessibility and cost effectiveness.

The following comments made from the participants also illustrate that MCEL could provide new learning experiences. In particular, students felt that they can take the learning on their own personal pace:

It is really useful to do answers in our own convenient time & it was new and attractive to do.

Sometimes, type in the correct format is not easy, when the wrong format sending, the cost has already happened, but the communication is no useful this moment.

Typing answers with mathematical operators took time. But the experience was good.

The survey results and our observation suggest that MCEL supports Bigg's constructive alignment in curriculum design (Biggs and Tang 2011), contextualized experimental learning, and personalized formative learning.

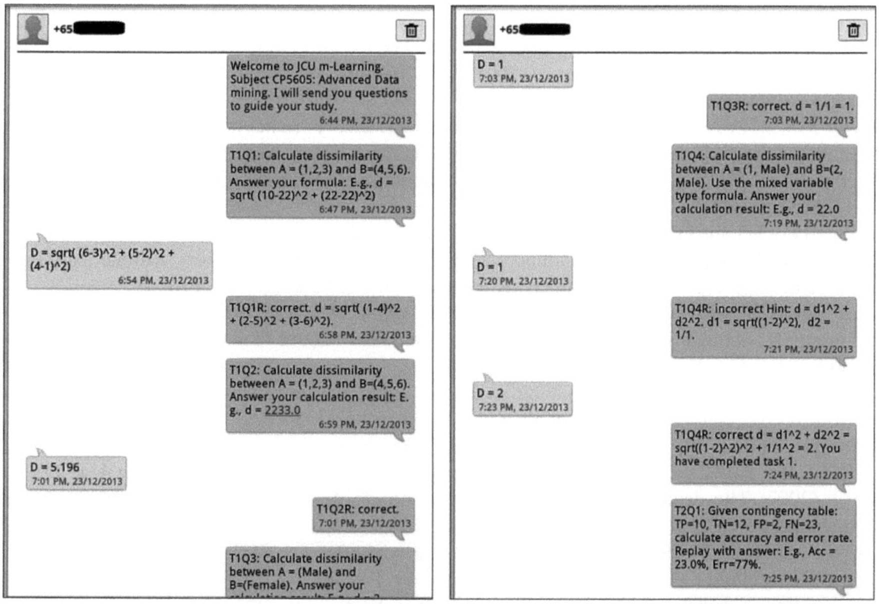

Fig. 13.3 MCEL interaction examples. Students interact with MCEL through SMS anywhere anytime. MCEL provides continues feedback on their progress

13.5 Conclusion

0We have identified a gap in the current literature and proposed a new approach for providing more engaging teaching and learning model: continuous automated formatives feedback extending beyond classrooms. The new model makes use of new emerging mobile technologies to provide continuous personalized formative assessments at much reduced cost. The proposed model can be supported with many available off-the-shelf systems allowing education institutions to create more engaging and vibrant learning environment for diverse student groups at low-cost. The mobile learning environment is even suitable for developing-countries, where no proper wireless network services are available. MCEL can be operated on conventional inexpensive GSM mobile services to provide personalized learning experiences.

References

Biggs J, Tang C (2011) Teaching for quality learning at university. Open University Press, UK

Boticki I, Wong LH, Looi CK (2013) Designing technology for content-independent collaborative mobile learning. IEEE Trans Learn Technol 6(1):14–24. doi:10.1109/tlt.2012.8

Brown AL (1992) Design experiments: theoretical and methodological challenges in creating complex interventions in classroom settings. J Learn Sci 2(2):141–178

Conole G (2004) E-learning: the hype and the reality. J Interact Media Educ 11
Coulby C, Hennessey S, Davies N, Fuller R (2011) The use of mobile technology for work-based assessment: the student experience. Br J Educ Technol 42(2):251–265. doi:10.1111/j.1467-8535.2009.01022.x
Devlin M (2012) Non-traditional student achievement: theory, policy and practice in Australian higher education. In: Proceedings of the 13th Pacific Rim first year in higher education conference 2010 on FYHE 2010: aspiration, access, achievement. Queensland University of Technology, pp 1–13
ITU (2011) ITU world telecommunication/ICT International Telecommunications Union
Kneebone R, Brenton H (2005) Training perioperative specialist practitioners. Mobile learning: a handbook for educators and trainers, Routledge, Milton Park, pp 106–115
Lefoe GE, Hedberg J (2006) Blending on and off campus: a tale of two cities
Manning CD, Schütze H (1999) Foundations of statistical natural language processing, vol 999. MIT Press, Cambridge
McInnis C, Hartley R (2002) Managing study and work: the impact of full-time study and paid work on the undergraduate experience in Australian universities
Ouyang C, Dumas M, Breutel S, ter Hofstede (2006) A translating standard process models to BPEL. In: Advanced information systems engineering. Springer, New York, pp 417–432
Parr T (2007) The definitive ANTLR reference: building domain-specific languages. Pragmatic Bookshelf
Roche E, Schabes Y (1997) Finite-state language processing. The MIT Press, Cambridge
Sharpe R, Benfield G, Roberts G, Francis R (2006) The undergraduate experience of blended e-learning: a review of UK literature and practice. Higher Education Academy, London
Song YJ, Wong LH, Looi CK (2012) Fostering personalized learning in science inquiry supported by mobile technologies. Etr&D-Educ Technol Res Dev 60(4):679–701. doi:10.1007/s11423-012-9245-6
Ting YL (2013) Using mobile technologies to create interwoven learning interactions: an intuitive design and its evaluation. Comput Edu 60(1):1–13. doi:10.1016/j.compedu.2012.07.004
Zhang B, Looi C-K, Seow P, Chia G, Wong L-H, Chen W, So H-J, Soloway E, Norris C (2010) Deconstructing and reconstructing: transforming primary science learning via a mobilized curriculum. Comput Educ 55(4):1504–1523. doi:http://dx.doi.org/10.1016/j.compedu.2010.06.016

Chapter 14
Application of ICT to Improve Rural Livelihood in Vietnam

Abstract Education and access to information are important factors for improving the livelihoods of rural communities and in the development of human resources in developing countries. Conventional approaches to education via schools and other government facilities, such as public libraries and post offices, are too slow to be adopted widely by communities. Recent progress in telecommunications technologies has now opened new opportunities to accelerate the process at a much lower cost. This has led to active research into community and telecentre based education systems. The World Bank, UN, and UNESCO have supported such projects in many developing countries, such as India, Bangladesh, and Vietnam in collaboration with their governments. However, many rural telecentre models have been conceived and tried but met with limited success. Many of the problems with these models have been traced to the lack of proper understanding of ownership, influence, native social patterns and their impacts. We reviewed and compared rural telecentre models that had been developed and studied over the last 10 years to better understand the key social concepts. A new model was proposed and two year pilot study was conducted. We report on the effectiveness and limitations of the new model.

Keywords ICT · Rural telecentre · Information service · Socio-economic systems · Socio-economic applications

14.1 Introduction

About 70 % of the world's people who make less than a dollar a day live in rural areas (World Bank 2008). Therefore, improving the livelihood of rural populations is considered crucial in improving the whole economy of many developing countries (Gollakota et al. 2012; de Janvry and Sadoulet 2010). Several studies have suggested that education and information provision are important factors for

J. Vong and I. Song, *Emerging Technologies for Emerging Markets*,
Topics in Intelligent Engineering and Informatics 11,
DOI 10.1007/978-981-287-347-7_14

improving the livelihoods of rural communities and in the development of human resources in developing countries (Colle 2007; Sachs 2005; Verspoor 1989; Lockheed and Verspoor 1991).

However, most developing countries do not have adequate resources to quickly improve the livelihood of rural areas, leaving many people in extreme hunger and poverty (Ahmed et al. 2007). Conventional approaches to education via schools and other government facilities, such as public libraries and post offices are too slow to be adopted widely by communities (Gomez et al. 2012). In addition, these approaches are too costly, requiring heavy infrastructure investment, such as for transport and adequate training and supervision of teachers. Recent progress in telecommunications technologies has now opened up new opportunities to accelerate the process at a much lower cost (Carvin and Surman 2006). A recent survey and pilot study by Gollakota et al. (2012) also suggested that ICT technology has benefited rural communities in enhancing knowledge of farmers about agriculture (e.g, crop cultivation).

Therefore, for over a decade, the World Bank, UN, and UNESCO have supported various telecentre projects in developing countries, such as India, Africa, and Vietnam, in collaboration with their governments to improve education and access to information in rural areas (Esterhuysen and Jensen 2001; Harris 2007; World Bank 2008). For instance, we conducted a two year pilot study of rural telecentre at communes in Vietnam under the framework of the project "Public Administration Reform in Ministry of Agriculture and Rural Development (MARD)" from 2006 to 2008. However, not all telecentre models in the last decade have been successful. Although the two years of implementation of information service models of the MARD project helped improve the livelihoods and increase incomes in rural areas, limitations of the current models were also observed.

Our analysis on existing information services and ICT programs suggests the main problems have been the lack of proper models of social concepts, particularly the following: ownership, influence (e.g., by peers), native social patterns and their impacts (Gomez and Camacho 2009; Gush 2008; Warschauer 2003). For instance, in a survey done in rural areas of Vietnam as part of our pilot study, the relevancy of product price information to local farmers was not as high as we thought. This was found to be due to the influences of traders, friends, relatives and neighbours. Figure 14.1 shows the ranking of the local people's need for information. The types of information are not exactly the kind of information that we would normally expect when considering the best practices of information providing models documented by Harris (2007) as shown in Table 14.1.

In this paper, we review rural telecentre models that have been developed and studied over the last 10 years and compare them to better understand critical social concepts that determine the success of ICT, telecentre, and information services programs in rural areas. A set of key social concepts were identified based on the best ICT and rural telecentre practices. A new information services model was designed based on the identified key social concepts. A two year pilot study of the new model was conducted in Vietnam using 13 rural telecentres, the web, internet

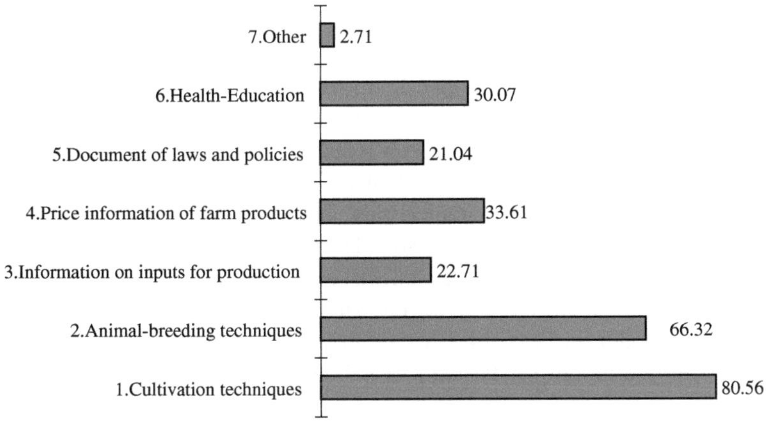

Fig. 14.1 The ranking of the farmers' need of information in Vietnam

Table 14.1 Categories of information that telecentres deliver

Category	Examples	Source	Delivery mechanism
Generic information	E-government, agricultural extension, distance education, e-commerce, e-health, news, weather, etc.	Government and national/regional institutions	Institutional partnerships
ICT-focused information services	E-mail, voice over IP, chat, internet searches, word processing, spreadsheets, presentations, ICT training, printing, photocopying, scanning, newsletters etc.	Local Telecentre owners, operators and staff	Creative and client-centric entrepreneurial activity at the telecentre
Information in support of development programs	HIV/AIDS awareness campaigns, micro-and small enterprise development support, skills training, micro-credit support, etc.	National/regional development initiatives	Pro-active marketing of Telecentres among NGOs, and INGOs
Locally-based information	Local laws, poverty reduction schemes, local NGO activities, yellow pages, job placement services, classified advertisements, market prices, bus schedules, doctors and vets lists etc.	Local activists, NGOs, community-based organizations, government offices, social entrepreneurs, volunteers, schools, etc.	Community outreach, community mobilisation, local networking

forums, and online databases of agriculture, policy, and market data. Results showed that operation of the telecentres were of practical use to farmers. 95.76 % of the farmers said the telecentres helped them increase their income.

The paper is organized as follows. Section 14.2 provides the background and context of the rural telecentre projects in Vietnam. In Sect. 14.3.1, we review major national programs for rural telecentres that have been implemented or planned in Asia and highlight common characteristics. In Sect. 14.3.2, we review various telecentre model developments and lessons learned from their success and failures. In Sect. 14.3.3, we then present key effectiveness criteria of information service models for farmers. In Sect. 14.4, we describe the two year pilot studies of new rural telecentre model. In Sect. 14.5, we describe the assessment results of the new information service model for farmers. We then conclude in Sect. 14.6.

14.2 Background

In year 2002, the emerging nation of Vietnam had at least 70 % rural population engaged in agriculture production but contributed to only 22 % of the GDP. Vietnam was facing the challenge of ensuring balanced social and economic development within the process of its international integration. It was expected that this process would have a direct impact on farmers, especially the poor. In order to manage current and forthcoming challenges in agriculture, Ministry of Agriculture and Rural Development (MARD) of Vietnam needed to make a commitment to implement effective public administration reform with a view to helping Vietnamese farmers take advantage of the processes of globalization as well as taking necessary measures against the challenges that might arise during the process; meeting the demands of the poor in rural areas.

Therefore, the UNDP had conducted two studies on farmer livelihoods in 2003 and 2004: Farmer Needs Study (UNDP VIE/98/004 2003; MARD 2003) and Feasibility Study on Rural Pilot Telecentres for Farming Households (UNDP VIE/02/016 2004) that revealed the plight of poor farmers and confirmed a need for information to increase the quantity and quality of agricultural production leading to higher incomes.

The main findings of the Farmers Need Study and Telecentre Feasibility Study are as follows. The Farmers Needs Study (2003), under the Public Administration Reform in MARD Project, identified problems faced by farmers in their efforts to increase the size and quality of agriculture production, and the alternative ways to support rural farming through public service delivery. The six-month Farmers Needs Study conducted not less than 1,200 door-to-door interviews in 60 communes in 30 randomly selected districts that were engaged in crop cultivation, animal husbandry, fishing and forestry. Among the key findings was that almost all farmers are poor, having household incomes of VND 10 Million or less. The farmers voiced their major problem as the lack of information to help them: control insects, rats and fungus devastation; and increase the selling price of agriculture products.

The 2004 ICT Feasibility Study took the form of a structured field survey of 20 randomly selected communes in 10 districts within 10 provinces across the nation, over a period of 4 weeks. The ICT Study found that information on agriculture output is most helpful to farmers; information is urgently needed for agriculture production reform; and technical information on agriculture extension, animal health and plant protection is urgent for farming communes. The ICT Study also highlighted that the district and communal extension at present has limited capacity to help farmers to resolve their problems and they themselves lack information and technical knowledge. The deployment of ICT for poverty reduction is already welcomed at three levels: commune, national and international.

The main findings of the two studies highlighted that the "information gap" lowers farm incomes and production due to:

1. Price exploitation of farmers by middlemen, slaughter houses, and food manufacturers;
2. Devastation of crops and animal husbandry by fungus, rats, insects and disease thus lowering production volume and quality;
3. Inability to find long-term buyers to stabilize their incomes;
4. Inability to seek the best credit sources (80 % of farming communes borrows to fund their agriculture production) thus paying higher interest rates and reducing incomes;
5. Inadequate knowledge of agriculture marketing skills to increase sales volume;
6. Lack of forward production planning thus lower incomes;
7. Inability to increase crop yields and husbandry;
8. Inadequate knowledge of agriculture extension to help farmers;
9. Outdated and non-specific training from agriculture extension;
10. Untimely and sporadic TV broadcast of agricultural information.

Following these studies and based on the positive results from the first period of reform in Vietnam, the MARD Macro Management Project in support of Public Administration Reform proceeded with assistance to MARD in rationalizing the institutional structure between the MARD and Departments of Agriculture and Rural Development. All the reform activities were being implemented with a clear objective of enhancing the capability to provide farmers with public services and better agricultural extension services.

This project paid special attention to the use of IT and media technology with the purpose of enhancing the capability of the MARD to liaise more effectively with the province level and commune level, specifically for the benefit of farmers who are its target customer. The project had built on the results arising from the first period of reform, specifically the five-year plan of Administration Reform which was made for MARD and an earlier comprehensive survey on farmers' needs in Vietnam (MARD 2003).

It was expected that this project would strengthen institutional capability and enhance the capability of officers who work at the MARD, especially in order to execute an activity plan on administration reform for the Ministry (in the period from 2006 to 2010). In this context, the project had a specific mission to assist MARD's administration reform efforts in several sectors which received preferential treatment in the activity plan.

UNDP had given its support to Vietnam in the Public Administration Reform from the very beginning. In addition, the Royal Netherlands Embassy in Hanoi had also supported this sector for some time, starting with pilot projects in Quang Binh Project (UNDP 1997) and in Hai Phong project (UNDP 1998). It was notable that UNDP acted as the coordinator to create favourable conditions for planning coordination, especially for the proposals for administration reform being made in close collaboration between the MARD and the Ministry of the Interior.

The objective of the project aimed at:

- Enhancing the capability of MARD in the rationalization of the organization structure at the central and regional levels with the purpose of meeting the demand for the provision of improved public services in agriculture.
- Enhancing institutional capability as well as that of officials working for key departments directly under the MARD in the implementation of the activity plan for administration reform in the period from 2006 to 2010.
- Assisting in the implementation of "Developing a strategy and applying IT technology in the State administration management" with the purpose of enhancing the effectiveness and prestige of the industry.
- Contributing to the strengthening of administration reform management capability by sharing information about solutions and necessary tools with other key administration agencies.

To avoid past mistakes of ICT, telecentre, and information service projects and to improve upon, a new information service model was needed. By reviewing information providing models serving for farmers in Vietnam, we identified that the main problems were the lack of proper models of social concepts, particularly the following: ownership, influence (e.g., by peers), native social patterns and their impacts according to our analysis on existing information services and ICT programs.

Before the establishment of telecentre model funded by Macro Reform Project-Ministry of Agriculture and Rural Development, there had been several information providing models mainly serving for farmers.

- Commune cultural post office is the project of Vietnam Posts and Telecommunications Group. Large-scaled system fully equipped by postal industry is considered the advantage of this model. The main objective of this model is to develop posts and telecommunications network to each commune, provide newspapers, magazines and connect the Internet. However, commune cultural post office places do not provide information on special demand of each commune or attract local people.
- Thanh Nua information centre takes the lead in the information centre project and is one of the development support programs sponsored by Action Aid Vietnam in Lai Chau Province. People visiting this centre for information must

fill in the Paper of Information Requirement covering contents such as name, address, date and required information. The centre staff search for the required information stored in the CDs. Clients can read the information on the computers or pay 200 VND for each printed paper. The centre also establishes groups of the same interest so that people can exchange experiences and knowledge together. However, operation of this centre does not closely link to agriculture—farmer support offices of Vietnamese government.

- VNCG is an electronic portal aiming at collecting, providing and exchanging information between state offices, schools and donors, which contributes to socio-economic development and hunger eradication and poverty reduction in Vietnam. VNCG provides information and knowledge, helping policy makers, businesses and ordinary people easily access. VNCG is connected to Global Electronic Portal to get information and share experiences with other countries in the world. One of the disadvantages of VNCG is not having a strong mechanism for receipt and reply of diversified information required by ordinary people. Therefore, effectiveness of VNCG is not high when considering the certain area with specific socio-economic conditions and information demand.
- Quotation system and market information is the component of Tea and Fruit Development Project with loans from Asian Development Bank (ADB) to invest in tea and fruit processing and commercial activities for improvement of farmers' income and agricultural product value as well. The objective of this system is said to be limited.
- www.agroviet.com.vn is one of the websites of Ministry of Agriculture and Rural Development which provides information of agriculture and rural development. This website is built up and maintained by computing centre ICARD (Agriculture and Rural Development Information Centre before). People accessing this website can send requirement/feedback. ICARD is responsible for replying. Sometimes, ICARD also receives some orders of foreign agricultural products. However, until now, there have not been any mechanisms for delivering these orders to relevant people. What ICARD can do is to collect information and post them on the website.
- Developing information system in the provinces is one of the activities in Trade Promotion Program—Ministry of Agriculture and Rural Development. Since 2002, the system of market information has been added to Trade Promotion Program—Ministry of Agriculture and Rural Development and expanded to 07 provinces including Lao Cai, Bac Giang, Nghe An, Dak Lak, Binh Dinh, Vinh Long and Can Tho. In these provinces, the program has provided money for training agricultural staff on skills of Internet using and market information collecting and processing. These staff implement the contract by visiting the markets or purchasing enterprises to collect the price market information of several specific products in each province. Then, the information is transferred to relevant people of Provincial Department of Agriculture and Rural Development, processed and posted on the website www.agroviet.gov.vn. Since October 2003, the market information system has changed according to Trade Promotion Program. The Program has limited the number of products needed

market price information to 1–2, they are fruit and green vegetables. Moreover, the Program has funded the systems for 09 provinces in the South. All of the collected information is delivered to ICARD and updated on the website www.agroviet.gov.vn, broadcasted on Vinh Long television then. Now, the system is operating efficiently.

In short, it could be said that there exist some basic disadvantages of information providing models serving for agriculture, farmers and rural area before telecentre foundation:

- These models provide information through a website; therefore, they do not meet continual demand on specific information of each locality, business or individual;
- Except commune cultural post office places, the system of facilities at each locality (commune level) is not fully equipped. Therefore, it is difficult to meet daily demand;
- There has not been a strong link between these models and the system of offices supporting agriculture—farmers at province and district levels yet.

Therefore, telecentres in the Macro Reform Project—Ministry of Agriculture and Rural Development were built on the spirit of learning strength and weakness of existing models. Particularly, on the extent of information and management, telecentres at commune work closely with local government. Furthermore, on the extent of meeting information demand, telecentres are located at commune so that they can provide specific information which is related to local demand.

Telecentre establishment was also based on successful experiences of telecentre network development in several countries with the same socio-economic conditions in rural area as Vietnam, for instance, India, Sri Lanka, Bangladesh, Philippines, etc.

14.3 Rural Telecentre Models

To avoid the past mistakes and improve upon, a new rural telecentre model was designed based on several past studies, by reviewing major national programs for rural telecentres, various telecentre model developments, and lessons learned from their success and failures.

14.3.1 National Programs for Rural Telecentres in Asia

The first community telecentre was opened in Vemdalen, Sweden, in 1985 and the concept rapidly spread to other European countries (Rose 1999), Africa (Fuchs 1997), Latin America (Galperin 2005), and Aisa (Best and Kumar 2008). UNESCO has been active in the Multipurpose Community Telecentre (MCT) area since 1996

(Rose 1999). Several countries in Asia are proceeding with national programs for telecentres, mostly aimed at delivering information and communication facilities to their rural populations who, for the most part, represent the majority of the poor people in the country. Such programs are under way in India, Nepal, Pakistan, Malaysia, Bangladesh, Sri Lanka, and the Philippines. This section briefly describes these.

14.3.1.1 India

"Mission 2007" (Swaminathan 2007)—Every Village a Knowledge Centre. Mission 2007 is a national alliance for access to ICTs for all India's 600,000 villages. It is targeting 100,000 Telecentres, each to serve six villages on average. The organizers of Mission 2007 recognize that it would be impossible to create a single, top-down telecentre model and deploy it nationwide. The success of the diverse initiatives in India has already proven that different communities require different solutions. Mission 2007 participants have agreed to accept their various differences and approaches; rather than argue one method is better than others, they plan to encourage villages to embrace the telecentre model most suited to their needs (Carvin and Surman 2006).

14.3.1.2 Nepal

From the same initiative, "Mission 2007" (Swaminathan 2007), the "Nepal-Mission Swabhiman" program was set up to target 1,000 Telecentres for 1,500 villages by 2010. The Tenth Five-Year Plan of the National Planning Commission of the Government of Nepal positions Telecentres as the medium for all development communication activities for the country. Mission Swabhimaan is a multi-stakeholder partnership of government ministries and agencies, international development agencies, donor agencies, non-government organizations, civil society organizations, academia, and private companies. It intends to create a knowledge revolution in rural Nepal to further meet the country's poverty reduction strategy and on the international level to meet the Millennium Development Goals.

14.3.1.3 Pakistan

The Government of Pakistan is planning for 10,000 Telecentres to be established in the rural areas of the country (Salman Ansari Technology Consultants (Pvt) Ltd 2006). The program intends to reduce the digital divide in Pakistan by extending low-cost computer hardware running localized Urdu language content to every village in Pakistan. This will help the fruits of technological advancements, starting with basic access to telecommunications, to reach poor and underprivileged communities across Pakistan, where the government has adopted a Universal Service Policy with the goal of preferably one Telecentre for every 5,000 people, or at least one Telecentre for every 10,000 people.

14.3.1.4 Malaysia

The government of Malaysia is implementing a program to install a Telecentre in all 927 mukims (local government districts) (Ministry of Rural Development 2001; Bashir et al. 2011). To ensure that the public has access to e-services provided by government, Telecentres have been set up in rural and urban areas by the Ministry of Rural and Regional Development, the Ministry of Energy, Water and Communication, the Ministry of Information, the State Government and the Local Authority. The Ministry of Rural and Regional Development has the Infodesa Project. The Ministry of Energy, Water and Communication is collaborating with Pos Malaysia, Telekom Malaysia, and the MSD group of companies (Kementerian Tenaga, Air dan Komunikasi) in setting up Rural Internet Centres (RICs). There are around 50 RICs throughout the country, all located at post offices where people congregate to pay utility bills and conduct other business. Members of the RIC can use the computer and Internet facilities for free. Each RIC has a website containing information on topics that are of interest to the local communities, such as health, agriculture, fisheries, tourism, arts and culture, security, entrepreneurship, local and national news, and local history. The RICs also conduct regular workshops to train the local community in the use of computers and the Internet.

14.3.1.5 Bangladesh

The government of Bangladesh has established "Mission 2011" to set up the Bangladesh Telecentre Network (BTN) (Raihan 2007) consisting of 40,000 Telecentres across the country by the 40th anniversary of the country's independence in 2011, so that every villager could have access to livelihood services through ICTs. The work is carried out by six task forces, responsible for Telecentre policy (advocacy), capacity building (peer-to-peer learning), locally relevant content development, technical support, resource mobilization and knowledge sharing. Each of the task forces will be led by an organization of repute with member organizations helping to implement the activities.

14.3.1.6 Sri Lanka

The Sri-Lankan government, with the support of the World Bank has established the E-Sri-Lanka program (ECDL Foundation 2002) that includes more than 350 Telecentres, known as Nenasalas, to disseminate knowledge and share information through the Internet, with the goals of poverty reduction, peace building, economic and social development and improving the IT literacy rate of the country. The government aimed to set up 1,000 Telecentres by the end of 2008. A majority of the Telecentres follow the community-based model, whereby the centres are established in a central place such as a temple, public library, or community hall. They provide services including high-speed Internet access, e-mail, telephone, computer training classes, and other ICT-related facilities. Content relevant to rural people is available in Sinhala and Tamil languages. Many of the centres are located at Buddhist temples.

14.3.1.7 Philippines

The government has a program for "Jumpstarting the e-Local Government Unit Project (e-LGU)" which includes 100 Telecentres (Teresa and Camba 2004). It is intended to enable LGUs to adopt computerization by empowering them with technology and applying the benefits of ICTs to local governance toward an improved quality of public service, increased transparency, greater accuracy and reduced processing time in local government transactions. This is expected to translate into better delivery of public services and to increase public access to information and services. The LGUs will establish interactive citizen-centric web presences and will set up community Telecentres. In summary, it is evident that each of these national programs displays some common characteristics:

1. Pilots: They all arose following pilot schemes (Caspary and O'Connor 2003; Dossani et al. 2005; Harris 2001). In some cases, these were government initiated; in other cases, they were initiated by civil society organizations and/or the private sector. In all cases, the pilots provided examples and learning of how to achieve desirable developmental outcomes with Telecentres. Moreover, the multiplicity of these pilots, combined with the accumulation of knowledge arising from the national programs that have sprung up from them, means that countries that have not yet proceeded to implementing a national program can learn from those that have and can, therefore, skip most of the pilot activity that the earlier countries went through.
2. Multi-sectoral partnerships: All the national programs consist of a mix of stakeholders, including, for instance; government ministries responsible for telecommunications, information, rural development, and agriculture, private sector organizations in the telecommunications industry, and civil society organizations working for poverty reduction and rural development. Often, the programs are supported by one or more international or bi-lateral development agencies.
3. High-level sponsor/champion: National programs for Telecentres usually draw on direct support from the highest level of government. This may entail the formation of a national body under the country's top leadership. Such an approach is required in order to achieve the level of cooperation, collaboration, and coordination of the various partners involved in bringing telecommunications and useful

information to the rural poor. It also serves to galvanize public interest in the program and to motivate participants toward a harmonized cause.

4. Co-ordination among government ministries: Under the high-level sponsor, it has always been necessary to create mechanisms to facilitate the correct level of inter-ministry coordination, which is never easy.
5. Multiple models of Telecentres: Against a background of diverse local conditions and disparities among the sponsoring organisations behind the pioneer Telecentre initiatives, most national Telecentre programs accept that there is no single model of Telecentre organisation to fit all situations. Accordingly, various models are being promoted simultaneously, with a mix of support structures for each.
6. Training staff: Large numbers of Telecentres require a large number of staff with specific skills, and formal training programs are necessary to achieve this. In the case of India, a dedicated national academy has been created to train Telecentre managers and operators.
7. Networking: National programs are promoted within networks of stakeholders as well as an overall Telecentre network, which may consist of regional or provincial networks. This facilitates sharing of resources and experiences, as well as the implementation of common standards for operational performance and service delivery.
8. International links: There now exist some international structures that foster Telecentre development, including national programs. These include the Global Knowledge Partnership, the Asia Telecentre Forum, and Telecentre.org. All the national programs have benefited from involvement with these institutions through opportunities for learning and sharing ideas.
9. Connected to national development goals: All national programs for Telecentres are driven by national development imperatives, sometimes stated as achieving the Millennium Development Goals, or as components of national poverty reduction strategies. Embedding Telecentre programs within such frameworks affords the essential validation of the programs as well as the criteria for assessing their impact.

As national Telecentre programs have all emerged from pilot projects, it is useful to emphasize that scaling up from a pilot project to a national program entails more than the mere replication of the pilot (Binswanger and Aiyar 2003; Alvord et al. 2004; Uvin et al. 2000). According to Telecentre.org, seven things are needed to scale up from pilot Telecentres to national Telecentre programs (Barbara Fillip and Foote 2007):

1. Flexible, responsive, and innovative social investment mechanisms to support the establishment of new Telecentres at the grassroots level.
2. Well-packaged, easy-to-replicate community services for Telecentres, such as telemedicine, remote learning, financial remittances, and e-government.
3. Simple, proven social enterprise models that Telecentres can use to generate community impact and financial revenue.

4. Flexible, ongoing training and support for hundreds of thousands of grassroots technology activists around the world.
5. Low-cost, easy-to-implement Telecentre technology platforms, including affordable and stable internet connections for rural areas.
6. Networks and partnerships that help good ideas travel far and wide—and help the Telecentre movement reach a global scale.
7. An enduring commitment to Telecentres and other grassroots technology initiatives from all sectors: governments, businesses, development agencies, and communities.

14.3.2 Telecentre Model Development and Global Lessons Learned

14.3.2.1 Rural Telecentre Models

Several Telecentre models have emerged over the past 10 year period of rapid growth around the developing world (Colle 2001). The chief characteristic of a Telecentre that determines the type of model it will follow is its ownership. Harris (2007) and Barbara Fillip and Foote (2007) defines five basic models for Telecentre ownership:

1. Civil society: the majority of Telecentres are non-profit, operated by Non-Government Organisations (NGOs). While they are quite often able to develop a close relationship with their community, funding is usually fragile and the scope for wider partnerships in support of a variety of information services is often limited.
2. Academic/research: Telecentres that take the form of action research.
3. Government: Telecentres that are initiated by a government or government agency, such as libraries, schools, and government agencies (ministry responsible for telecommunications, or it may be a ministry of agriculture, information or rural development or some other specialised ICT agency). They may or may not aim to achieve sustainability/profitability on an individual Telecentre level.
4. Private sector: For-profit Telecentres (e.g., (Kiri and Menon 2006; Kuriyan et al. 2006)), such as Phone shops, Public Call Offices (PCOs), and corporations to assist their primary suppliers (e.g. providing agricultural support). Agricultural activities are found most often in for-profit Telecentres.
5. Hybrid franchise model, involving two or more of these: These are set up to scale up to large numbers of Telecentres.

14.3.2.2 Applicability of the Various Models of Telecentres

Other than ownership, the choice of Telecentre model depends on other factors, including local circumstances (Graziella Cardoso Bonadia et al. 2008), particularly the extent to which the host community is able to support the centre financially. The poorest locations, often the most remote, will have the least resources to support a Telecentre, but may have the highest claim for the benefits that a Telecentre could offer. Accordingly, Telecentre models sit on a spectrum from wholly commercial at one end to wholly subsidized at the other (Barbara Fillip and Foote 2007). In most cases, operational Telecentres are hybrids that sit somewhere in between these opposites. One consideration which often determines where Telecentres are located between these ends of a spectrum is the extent to which telecommunication services and connectivity are considered to be a public good that merits government funding or a private service that should be provided by the market.

Telecentres that can be classified along this spectrum exhibit differing characteristics based on how close they are to each end of the spectrum. Those that are closer to the market-supported model are likely to remain financially healthy in the longer term, but they tend to focus more on revenues than on public services. On the other hand, Telecentres that are more socially oriented usually exhibit financial weaknesses that make their long term existence more questionable. In some countries, it can be observed that the Telecentres that began in positions more towards the outer ends of this spectrum are moving closer towards the middle, and, as their numbers increase, the main task that has emerged is to manage Telecentre programs in such a way that they are able to deliver on both sides of the equation, that is, maintaining robust finances at the same time as delivering essential public services to their host communities. The term social enterprise model is emerging to describe this form of Telecentre operation.

14.3.2.3 Best Practices of Rural Telecentre Operation

Despite the variety of organisational models for Telecentres, experiences over the years have given to rise to a few best practices that are found to be more or less universal. Barbara Fillip and Foote (2007), Harris (2007) identified twelve best practices as follows:

1. Development orientation (e.g., a poverty alleviation focus).
2. Local community involvement in design and implementation: the community should be the primary stakeholder.
3. Marketing and community engagement: must market its services among its intended audience if it wishes to make an impact on local development.
4. Accessible premises: location greatly affects accessibility and the use of facilities in Telecentres.

5. Partnerships and Networking: involve all stakeholders in the community: community leaders, community organizations, schools, health centres, agricultural extension agents and input suppliers, and local cooperatives.
6. Robust and appropriate technology solutions should be driven by the decisions made about what type of information services it will provide.
7. Locally relevant content and applications (the information) to the community.
8. Enabling ICT policy and regulatory environment scaling up to a more widespread deployment of Telecentres.
9. Capacity building of the people deploying and maintaining it and Telecentre users (e.g., the community leaders).
10. Role of local champions and telecentre/kiosk operators for sustainable management of telecentres.
11. Business/Sustainability plan to raise funds and obtain other kinds of support needed to start and sustain the operation.
12. Continuing evaluation involving all stakeholders for policy adjustment and to implement improvements.

14.3.3 New Rural-Telecentre Model Based on the Social Model

The characteristics of telecentre ownership models, best practices, and national programs are the accumulation of the decades of knowledge that have shaped ICT model development and planning in Asia. Therefore, they form the basis of the effectiveness criteria of rural telecentre models. Table 14.2 lays out the key effectiveness criteria of rural telecentre models. These were then used to design, implement, and evaluate a new model for providing information to farmers based on their needs in the pilot of the MARD telecentre project at communes in Vietnam from 2006 to 2008.

14.4 Two Year Pilot Study of the Social Rural-Telecentre Model

The pilot of 13 rural telecentres (RTs) at commune level was designed as a sub component of the "Support to Administration Reform at MARD" Project and later a sub component in the MARD Macro Management—Support to the Administration Reform (2006–2010).

Table 14.2 The key effectiveness criteria of rural telecentre models based on telecentre ownership models, best practices and national programmes

Category	Criteria
Telecentre ownership model	The ownership model is positioned on a spectrum where it is wholly commercial at one end and wholly subsidised at the other
Telecentre best practices	Development orientation Information focus Local community involvement in design and implementation Marketing and community engagement Accessible premises Partnerships and Networking Robust and appropriate technology solutions Locally relevant content and applications Enabling ICT policy and regulatory environment Capacity building Role of local champions and telecentre/kiosk operators. Business/Sustainability plan Continuing evaluation
National telecentre programme	Pilots Multi-sectoral partnerships High-level sponsor/champion Co-ordination among government ministries Multiple models of telecentre Training staff Networking International links Connected to national development goals

14.4.1 Project Objective

The pilot's objective was to identify "an appropriate model which was technically and financially sustainable and capable to provide farmers with information supporting production development and income improvement". The specific outputs of the sub-component include: (i) the current situation of communes participating in the pilot reviewed; (ii) 13 RTs at commune level in 10 provinces belonging to 8 eco-agricultural economic zones established; (iii) enhanced capabilities of local staff participating in the pilot; (iv) information resources developed; (v) information channels and dissemination forms developed (vi) results, impacts of the pilot assessed and proposals for replication. During the 3 years pilot period, 13 RTs came into sustainable operation and introduced benefits for the local community.

By the end of pilot period, the Project Management Unit (PMU) decided to recruit a local contractor to carry out a final external assessment of the initial results and impacts of the pilot. This assessment was under the supervision of an international consultant who was experienced in RTs in the region and worldwide. After being provided with facilities and enhanced institutional capability, the RTs have come into operation and have accomplished significant results over the two year pilot period.

14.4.2 *Facilities of Rural Telecentres (RTs)*

Arising from the survey it can be seen that the facilities in most of these centres were comparatively good.

14.4.2.1 Office Centre

All the centres were spacious, commodious, near the market, and convenient for people to use the service. Some centres were arranged separately, independently from the People's Committee headquarters. These centres were usually located in the centre of cultural communities. Due to the limited condition of the commune, some centres have placed their offices in the People's Committee head quarters, the area of which was not as large as that of the centre Thanh Binh, Tien Giang Province (about 15 m^2).

However, in some of these villages whose centres were considered as part of the People's Committee, their office centre in the People's Committee were relatively large, for example, the centre in Dong Tho village, Thai Binh province was approximately 50 m^2 in area. For the centre of Tan Lap village, Binh Phuoc province, a part of the office centre with 40 m^2 in area, in order to serve the needs of studying the computing of cadres as well as people in the village, the People's Committee has arranged a classroom which is available to set up 30 computer units to meet the purpose of negotiating between the centre and the Farmers to open a computer training class.

These centres had a large area which serves the needs of the people with their demand for accessing computers, reading materials and newspapers. From the practical implementation, setting the centres in the People's Committee is not suggested because this might be intimidating for some people. However, after the period of pilot operation, these centres were considered to be a part of the People's Committee and the offices located in the People's Committee are working well.

14.4.2.2 Machinery, Equipment

In addition to the initial machines equipped in the project, the centre has so far owned the computer systems with high-speed internet connection (ADSL) which worked so well and each centre was complemented further 02-03 computers.

According to officers in the centres, during the operation they may sometimes encounter problems with machines and software, but these also have been processed in time to serve the best needs of the people. Compared to the time of establishment, the configuration of the computer at the centre is high enough to meet the requirements of the activities of the centre. However, with the development of information technology at present, the computer centres must be upgraded.

14.4.2.3 Organizational Structure During the Implementation of Project

There has been participation at all levels from central, province, district and commune level. At central level, the Central PMU was established to manage the whole RT nationwide. There also has been participation of officers from a number of key Departments in organizing and implementing the project, including: Central Coordination Group (including 09 officers from Departments, Centres, heads of Division under Departments), Central Support Group (comprising of 6 officers at Division level, experts from units under MARD: MARD Office, Department of Science and Technology, Department of Planning, National Agricultural Extension Centre, Informatics and Statistics Centre) and groups of coordinating officers of the Ministry (civil servant from departments, institutions, schools, centres under MARD). The participation of officers from MARD played a significant role in the success of the project, coordinating RTs at various geographical regions and areas, unifying activities of RTs, as well as creating a seamless communication channel from Central to unit. National Execution Procedures have benefits during the implementation of the procedures such as planning, preparing financial reports, progress reports, training for officers. However, in a bid to spreading the project, central levels must concentrate on issuing guiding documents at macro level, and detailing regulations and forms for all RTs nationwide. Simultaneously, it should issue regulations to encourage the development of centralized information sources (such as the establishment of a national data base on Agricultural and Rural Development), which would result in easier accessibility to information for farmers.

The organizational structure during the implementation of the project is illustrated in Fig. 14.2.

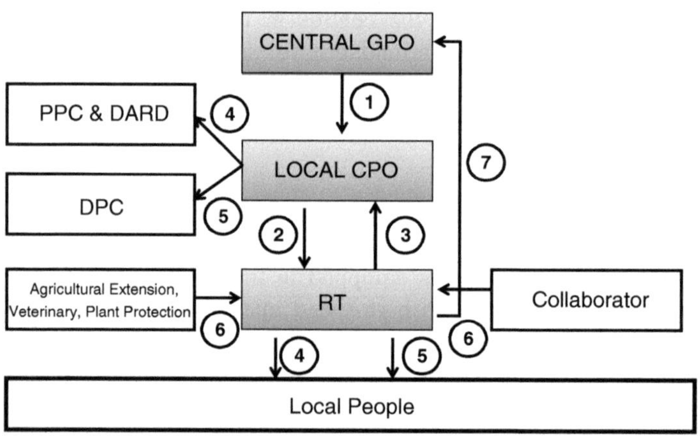

Fig. 14.2 *1* Directions of central PMU, *2* directions of regional PMU, *3* operation reports of RTs to regional PMU, *4* regional PMU report to department and provincial people's committee, *5* regional PMU report to district's people committee, *6* cooperation of RTs with agricultural extension and plant protection stations, as well as collaborators, *7* financial reports of RTs to central PMU, *8* provision of information to farmers

As shown in Fig. 14.2, the level of interest varied among the provinces or to be more exact, among individuals. The officers above concurrently hold many positions whereas the obligations of each position are not well defined. Therefore, for further implementation, the responsibilities of provincial and district level officers must be clearly defined in order to optimize current resources of RTs, especially in terms of providing information resources to farmers.

14.4.3 Assessment Method

In order to draw lessons learnt for effective, sustainable and actionable recommendations for replication of the model, a final external assessment of the results and impacts of the RTs was conducted based on the guidelines of (Whyte 2000; Harris 2007; Barbara Fillip and Foote 2007) and the initial assessments (MARD 2003). Specifically, the assessment objectives to be addressed were:

- To assess the appropriateness and performance (relevance and impact) of the RTs against the expectations specified in the project documents.
- To draw lessons learnt from these pilot RTs; or/and;
- To make specific and actionable recommendations with a view to bringing into play the stronger points and limiting the weak points, thereby facilitating a wider replication of the mode in broader manner.

Basically, the key objective of the Assessment is (1) to take into consideration appropriateness of RTs model after 3 years pilot period and (2) assess the results and impact of the RTs in supporting production development and income improvement for local farmers (3) to make specific and actionable recommendations for a wider replication of the model.

In reality, it has been a quite a long time since the design, building and implementation of such RTs were carried out. Therefore, it is essential to carry out such an assessment of the appropriateness of the RT model. The local contractor must assess the appropriateness of this model in accordance with (1) policy and incentives of Vietnam Government and sponsors and with (ii) the current situation in rural Vietnam.

Additionally, it is vital to consider the existing results and impacts of the RTs; among the fundamental issues to be researched by the independent consultant is: what is the level of effectiveness of the project objectives, the benefits that local farmers enjoy, and the positive changes after the implementation of this project etc.

Factors that influence both negatively and positively the sustainability of the pilot model were specifically researched. Furthermore, it was required to perfect a model for wider replication in the future by making an evaluation and analysis on the institutional and financial aspects, and the participation of competent authorities at different levels and the local people to ensure the long term existence and effectiveness of the RTs.

14.5 Assessment Results of the New Telecentre Model

At the completion of the two years of the pilot project, a survey was conducted to see the changes in the ways that farmers searched for information, before and after the establishment of the rural telecentre (RT). Figure 14.3 shows the numbers of clients directly coming to RTs for information collection by genders, up to the middle of September 2007. Figure 14.4 shows that due to the convenience of the RT, the previous ways of accessing information, such as by word of mouth, decreased significantly. Moreover, the local people became more active by going to the RT to access the internet, read materials, and participate in the agricultural technical trainings held by the RT's staff with agriculture officers and the farmers' association.

According to data from the survey, 95.76 % people said the RT has helped them increase their income as shown in Fig. 14.5. The improved income may vary but generally the RT has had a practical effect on people. Thanks to the information provided by the Centre Office, farmers can apply the experiences and technical science to enhance cultivation and production productivity as well as avoid the negative consequences caused by natural disasters and epidemics.

The data also shows that there are 65.69 and 29.93 % of those questioned feel satisfied or very satisfied respectively with the quality of the services provided by the RT as shown in Fig. 14.6. Only a small number of people are still not satisfied with the service quality of the centre (2.64 %). Explaining their opinions of feeling less satisfied, the surveyed subjects indicated the following reasons:

1. As for localities in the process of restructuring agriculture sector, especially those re-organizing their crops and raising animals, farms and fast developing

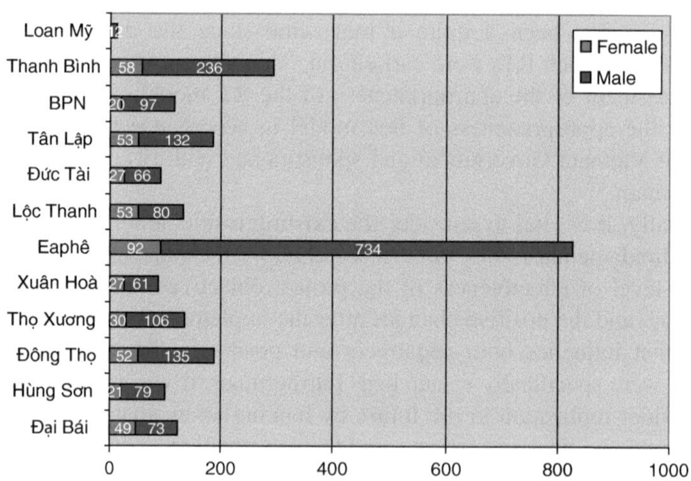

Fig. 14.3 Numbers of clients directly coming to RTs for information collection by genders for the first 6 months since its full operation

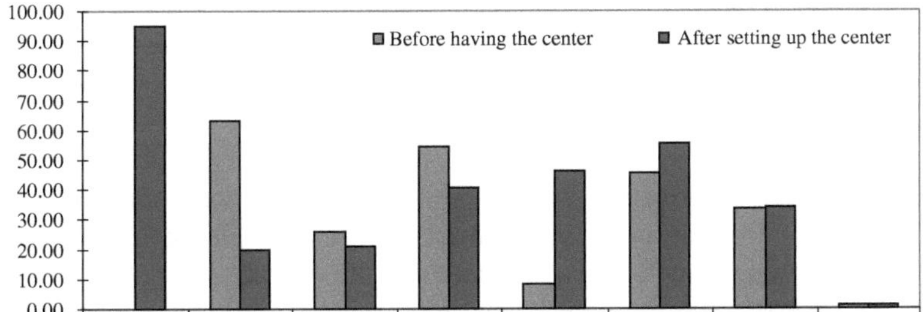

	1.From the Telecenter	2.By word of mouth	3.From newspaper	4.Radio/T.V at home	5.From internet	6.Officials from center	7.Officials from local	8.Other
Before having the center		63.19	25.63	54.38	8.19	45.42	33.47	0.97
After setting up the center	95.21	20.00	20.76	40.42	46.04	55.69	33.96	1.18

Fig. 14.4 The ways to search information before and after the establishment of rural telecenter

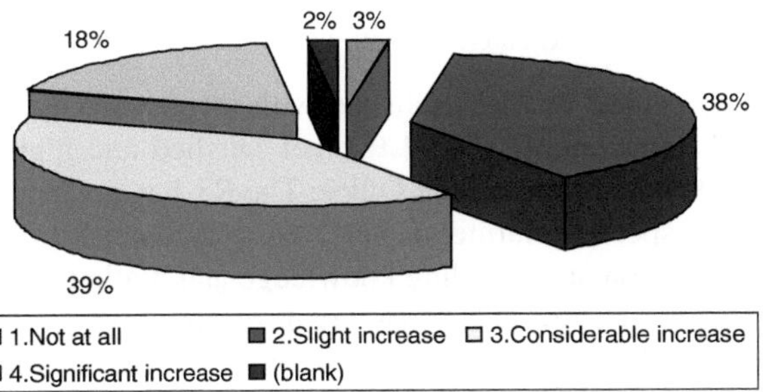

Fig. 14.5 Percentage of farmers having income improvement with support of rural telecentres

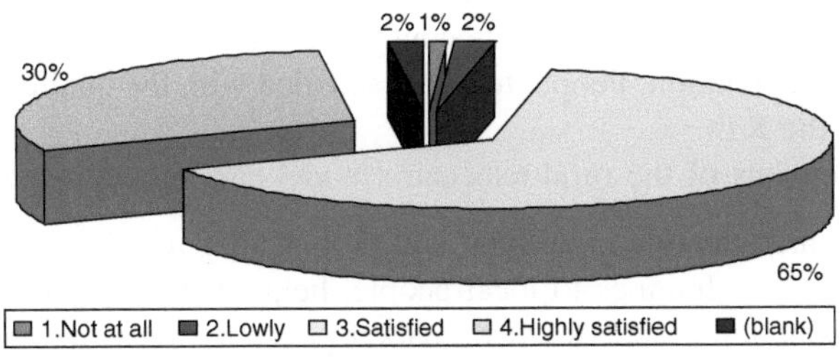

Fig. 14.6 Level of satisfaction of farmers with services at rural telecentre

farm households (Dong Tho Commune, My Loan), there is a great demand for information for the farmer. In fact, whilst sources of information provided by the Centre Office are diversified, they did not match with the actual needs for information for the farmers. Therefore, the information provided by the Central Office needs to be in more detail, making it easier for people to understand and apply.

2. Although the strength of the RT is the usage of a network connection, allowing access to information, the number of the accessed internet addresses is small compared with the available online resources. The Centre Office focuses only on a few websites such as www.mard.gov.vn; www.khuyennong.gov.vn; www.agiviet.com.vn to obtain information on production and animal husbandry. As a result, the information sources are limited and monotonous. There was little general information regarding market prices, people health and policies of the state compare with the actual needs of the people;
3. Some RTs have not coordinated closely with the local agencies and organizations, so its information sources are limited. The announcement of policies and guidelines of the local government reached the farmers more slowly than other information channels (such as local radio);
4. Information of the Centre Office has not been sufficient and unable to meet the information demand of the farmers, especially for poor people who not able to access other information channels. The Centre Office should print more leaflets on various information fields in order to provide more information to farmers and at the same time update information to catch up with actual demand in a timely manner.

Although some people feel less satisfied with the service quality of the Centre Office, in general, most people (96.62 %) feel satisfied and highly appreciate the operational effectiveness of the Centre Office. The RT has practical meaning for the life of the people (especially farmers). Since the establishment of the RTs, people gain technical cultivation and breeding knowledge and market prices more quickly. At the same time, the Centre Office has organized training on computer skills to local people. Before the RT was established, farmers were unable to find answers for their questions, but since the RT has come into operation, the needs of the people have been met, although not completely, but compared with the reality, people have partly feel satisfied, especially, the RTs also had a positive influence to people of some neighbouring communes in the region.

The RTs have made a considerable contribution to the development of the local socio—economic situation. People felt very satisfied with the quality of information received from the RTs.

The achievements of the rural telecentre model can be summarized as:

- Telecentres are effective in seeking and propagating information on production and cultivation techniques to local people, helping them gradually change traditional production practices. Also, market prices are provided, helping farmers avoid being forced in terms of price and minimizing risks incurred by epidemics calamity and market changes;
- Telecentres help local people update their socio-economic, education and legal information and contributing to their improved awareness and livelihoods;
- Telecentres are successful in changing the awareness of local people in project communes, showing them the usefulness and effectiveness of information technology to normal life. Many families have purchased computers and connected them to the internet and shared information with telecentres and the neighbourhood;

- Telecentres form a 2–way information channel that both provide information on production techniques to improve production conditions and increase income, while collecting responses and feedbacks of farmers to relevant authorities. Telecentres are a really useful tool for local authorities in administrative management;
- Telecentres have offered several services which help develop information knowledge and awareness in rural areas (informatics training and internet access etc.), creating opportunities for rural young labourers to get access to new knowledge, and training informatics for staff of local authority staff which contributes to improved administrative reform at communes.

14.6 Conclusion

Access to information technology is vital to improving the livelihoods of rural communities (Badshah et al. 2005; Proenza 2001). It can help save time in the long run and save costs, benefiting not only the agriculture sector, but also the economy of the whole nation, increasing incomes and production efficiency for farmers and urban dwellers. However, to be successful, ICT projects for community support require deeper understanding of social concepts. Earlier models, such as commune cultural post office, information centre, electronic portal, quotation system, and market information, mainly focused on providing information through a website. These earlier models did not meet continual demands for specific information from each locality, business or individual. This can all be traced back to some aspects of misunderstanding of the interactions between users and computing systems. The key effectiveness criteria of information service models were aimed at solving these problems and at developing a new model for providing information to farmers in the MARD telecentre project. The results show that the new model was effective in guiding project managers to develop an ICT system for farmers.

After 2 pilot years, the rural telecentre model has not only been successful in providing market and production technique information to producers, but also helped to improve awareness of leaders and the community on benefits of information in the economic integration period. Results of the review show that operation of telecentres is of practical use to farmers (62.50 % of those interviewed said telecentres are very useful). More importantly, the new information service helped 95.76 % of the farmers increase their income. This strongly indicates that the new information service model was effective in addressing the shortcoming of the previous approaches.

Despite the above mentioned major achievements, the telecentre model also has several limitations. Under Vietnam's current circumstances, scaling up of this model to a national scope necessitates some conditions which are not universal to rural areas in Vietnam. Disadvantages of the telecentre model are identified as:

In financial terms, it is obvious that current telecentres, including those with very good operation like Dong Tho's and Eaphe's, for the time being, cannot live on their own. This is mostly because socio-economic and education conditions in rural Vietnam are not developed enough for fee based services to be provided. Also, the financial arrangements for maintaining the current telecentres in the coming years are not an easy task.

In terms of staff, the current workforce of telecentres are working quite well because they have been selected and trained thoroughly. However, maintaining this force in the coming times, especially when the project ends its financial support, will be difficult.

Regarding institutional arrangements, in this model, the commune level are showing greatest enthusiasm and dedication. The role of the district level remains unclear. In the structure of the local project management unit the provincial level has only one member. This official, in most cases, is only a staff of the provincial Department of Agriculture and Rural Development. But in reality, the provincial level has a very critical role in co-ordinating and arranging responses to questions from telecentres and in arranging budgets for telecentres from now on. In the whole system, this is the most "vulnerable" link which needs consolidating.

At central level, the role of the central level support team remains limited. The forum in MARD's website, in theory, could have been an important support to the telecentres. However, the forum has failed to utilize its advantages as there have been very few discussions so far. This is because of the lack of a mechanism with commitments and monitoring strong enough for members of the team to increase their spirit of responsibility and self-awareness in their mandate to work for telecentres.

Among the limitations, sustainability is still the main remaining issues as in previous projects (Best and Kumar 2008; Mayanja 2001; Proenza 2001). However, the rapid penetration of cell-phone networks in developing countries (Badlani 2012) is closing this gap fast.

Acknowledgment This paper is based on the fieldwork and research conducted specifically in the Rural Telecenter Project under the framework of the Public Administration Reform (PAR) Project at the Ministry of Agriculture and Rural Development of Vietnam (VIE/02/016), funded by the Government of Vietnam, Royal Government of Netherlands, UNDP and supported by Intel USA, where John Vong was the UNDP Senior Technical Consultant. We would like to thank UNDP (United Nations Development Program) and Royal Netherlands Embassy for the grants and support. We are truly grateful to the Department of Agriculture and Rural Development, District People's Committee, Ministry of Agriculture and Rural Development, Provincial People's Committee, and all villagers in Vietnam for their participation and support in the project.

We would like to thank UNDP, in particular Jordan Ryan, Subinay Nandy, Steven Gieger, Jonas Lovkrona, Ninh Thi Hoa, and Le Le Lan. The Government of Vietnam, Ministry of Agriculture and Rural Development, provided strong national leadership especially under Vice Minister Nguyen Van Dang and Project Manager Nguyen Duc Son, without which the success of the Rural Telecentre Project may be dampened. The baseline study conducted by the Centre for Water Resources Economic Research and Centre for Water Resources Software was important prior to the start of the Rural Telecentre pilot. The final evaluation of the Rural Telecentre pilot under research consultant Roger Harris was fair and accurate. The strong support of the Government of Royal Netherlands, in particular Ambassador Gerben de Jong, provided impetus to the overall framework of the PAR in MARD Project. Lastly, the opinions and findings expressed herein are those of the authors alone.

References

Ahmed AU, Hill RV, Smith LC, Frankenberger T, Gulati K, Wiesmann DM, Quabili W, Yohannes Y (2007) The world's most deprived: characteristics and causes of extreme poverty and hunger, vol 43. International Food Policy Research Institute, Washington, DC

Alvord SH, Brown LD, Letts CW (2004) Social entrepreneurship and societal transformation an exploratory study. J Appl Behav Sci 40(3):260–282

Badlani D (2012) Africa: opportunities and challenges in a growing economy. http://www.roycefunds.com/news/global/2012/0125-africa-opportunities-challenges-growing-economy.asp

Badshah A, Khan S, Garrido M (2005) Connected for development: information kiosks and sustainability, vol 4. United Nations Publications, New York City

Barbara Fillip, Foote D (2007) Making the connection: scaling telecenters for development. Washington, DC

Bashir MS, Samah BA, Emby Z, Shaffril HM (2011) Impact of individual characteristics and telecenter success in Malaysia. Aust J Basic Appl Sci 5(9):371–380

Best ML, Kumar R(2008) Sustainability failures of rural telecenters: challenges from the Sustainable Access in Rural India (SARI) project. Info Technol Int Dev 4(4):31-45

Binswanger HP, Aiyar SSA (2003) Scaling up community-driven development: theoretical underpinnings and program design implications, vol 3039. World Bank Publications

Bonadia GC, Ávila IMA, Ogushi CM, de Holanda GM (2008) Modeling the acceptance and use of telecenters in Brazil. J Technol Manag Innovation 2(4):86–97

Carvin A, Surman M (2006) From the ground up—the evolution of the telecentre movement. In: Andy Carvin, Surman M (eds)

Caspary G, O'Connor D (2003) Providing low-cost information technology access to rural communities in developing countries: what works? what pays? Citeseer

Colle R (2007) Communication shops and telecenters in developing nations. In: Gurstein M (ed) Community informatics: enabling communities with information and communication technologies. Idea Group, Hershey, pp 415–445

Colle RD (2001) ICTs, telecenters and community development. Cornell University, Ithaca

de Janvry A, Sadoulet E (2010) Agricultural growth and poverty reduction: additional evidence. World Bank Res Observer 25(1):1–20

Dossani R, Misra D, Jhaveri R (2005) Enabling ICT for rural India. Asia Pacific Research Center, Stanford University and National Informatics Centre. Last Accessed 14 July 2008

ECDL Foundation (2002) e-SRI LANKA PROJECT: Using ICT to build capacity and foster social integration

Esterhuysen A, Jensen M (2001) The community telecentre cookbook for Africa

Fuchs R (1997) If you have a lemon, make lemonade: a guide to the start-up of the African multipurpose community telecentre pilot projects. International Development Research Centre (IDRC). Available at www.idrc.ca/acacia/engine

Galperin H (2005) Wireless networks and rural development: opportunities for Latin America. Inf Technol Int Dev 2(3):47–56

Gollakota K, Pick JB, Sathyapriya P (2012) Using technology to alleviate poverty: use and acceptance of telecenters in rural India. Info Technol Dev 18(3):185–208

Gomez R, Pather S, Dosono B (2012) Public access computing in South Africa: old lessons and new challenges. Electro J Info Syst Dev Ctries 52(1)

Gomez R, Camacho K. (2009) Who uses ICT at public access centers? age, education, gender and income differences in users of public access to ICT in 25 developing countries. Paper presented at the second annual SIG GlobDev workshop, Phoenix, Arizona

Gush K (2008) Towards a more personalised user experience and better demographic data on the digital doorway public computer terminals. Paper presented at the 5th Prato community informatics and development informatics conference: ICTs for social inclusion: what is the reality? Prato, Italy

Harris R (2001) Telecentres in rural Asia: towards a success model. Europe 40 (23.4):13–17
Harris R (2007) A framework for designing telecentres—UN-APCICT. UN, APCICT-ESCAP
Kiri K, Menon D (2006) For profit rural kiosk in India: achievements and challenges. i4d Magazine
Kuriyan R, Toyama K, Ray I (2006) Integrating social development and financial sustainability: the challenges of rural computer kiosks in kerala. In: International conference on Information and communication technologies and development, 2006 (ICTD'06). IEEE, pp 121–130
Lockheed ME, Verspoor AM (1991) Improving primary education in developing countries. Oxford University Press for World Bank
MARD MOAARD (2003) Farmers needs study. VIE/98/004/B/01/99
Mayanja M (2001) The african community telecenters: in search of sustainability. ICT for Education Programme
Ministry of Rural Development (2001) Program Infodesa. Ministry of Rural Development. http://www.kejora.gov.my/en/web/guest/program_infodesa. Accessed Dec 2012
Proenza FJ (2001) Telecenter sustainability: myths and opportunities. J Dev Commun 12(2):94–109
Raihan A (2007) Bangladesh telecentre network (BTN). IDRC Canada, Bangladesh
Rose JB (1999) Multipurpose Community telecentres in support of people-centred development. In: Development CNoITf (ed) Information technology and globalisation: implication for people-centred development. SFI Publishing, London
Sachs J (2005) The end of poverty. Penguin Press, New York
Salman Ansari Technology Consultants (Pvt) Ltd (2006) Telecenters and community resource and information centers in Pakistan
Swaminathan MS (2007) Grameen Gyan Abhiyan (rural knowledge movement). M.S. Swaminathan Research Foundation. http://www.mission2007.in/. Accessed December 2012
Teresa M, Camba M (2004) Jumpstarting electronic governance in local government units in the Philippines (eLGU). From policy to reality. In: Proceedings of International Conference on e-Governance ICEG 2004
UNDP (1998) Public administration reform programme, Hai Phong city pilot project. United Nations Development Programme
UNDP (1997) Strengthening capacity for hunger eradication and poverty reduction in Quang Binh province, vol VIE/96/007. United Nations Development Programme
Uvin P, Jain PS, Brown LD (2000) Think large and act small: toward a new paradigm for NGO scaling up. World Dev 28(8):1409–1419
Verspoor A (1989) Pathways to change: improving the quality of education in developing countries. World Bank discussion papers 53. ERIC
Warschauer M (2003) Technology and social inclusion: rethinking the digital divide. MIT Press, Cambridge
Whyte A (2000) Assessing community telecenters: guidelines for researchers. International Development Research Centre (IDRC), National Library of Canada
World Bank (2008) World development report—agriculture for development. World Bank, Washington, DC

Conclusion

Over the last 25 years, the social and development economists, development specialists, bankers, fast moving consumer goods (FMCG) corporations and political scientists from the developed world have set their eyes on the emerging markets. Their focus is based on economics, international development, business, or political reasons. While some of these observers may have despised the rise of emerging markets, others have embraced it, but none has ignored it.

Given that more than half of the world population resides in the BRIICS, one could not help but wonder what is the next wave of consumer technologies coming to these countries? And how these countries will influence the inventions and innovations of the next wave of technologies? Who would have thought that Alibaba.com from China would be the world's largest ever initial public offering of US$21.8 billion? Some commenters have described their business model as one that combines Paypal, eBay, Amazon and Google. Do you know that, just behind the US, Indonesia is the country with the second largest number of Facebook users, and followed by India? From this, we can see that the world is leaning towards the emerging markets.

The authors have little doubt that the three pillars of technological growth are health, education, and financial services. Co-incidentally these three pillars form parts of the millennium development challenges. This means that development grants for research and capacity building in health and hospitals, education and schools, and financial services and payments are forthcoming. This, in turn, leads to increased consumption in the emerging markets. Also, the trajectory growth of new technologies is determined by the rise of the middle class.

The policymakers of both developed and the developing countries should be aware of the technology trends and their interdependence with the emerging markets. Technology skills and knowledge should be taught at all education levels. Teachers, medical doctors and nurses, and bankers should be tech savvy to keep up with the technological changes at their workplace. With the use of mobility solutions from banking to health to education, it is high time the costs of financial services, health services and education services be reduced.

J. Vong and I. Song, *Emerging Technologies for Emerging Markets*,
Topics in Intelligent Engineering and Informatics 11,
DOI 10.1007/978-981-287-347-7

The world is ever-changing in an unpredictable way amidst many strands of market forces at play. Some of these change patterns are obvious while some are subtle. This book presents a glimpse of the emerging markets driving, and being driven by, emerging technologies.

MIX
Papier aus verantwortungsvollen Quellen
Paper from responsible sources
FSC® C105338

If you have any concerns about our products,
you can contact us on
ProductSafety@springernature.com

In case Publisher is established outside the EU,
the EU authorized representative is:
Springer Nature Customer Service Center GmbH
Europaplatz 3, 69115 Heidelberg, Germany

Printed by Libri Plureos GmbH
in Hamburg, Germany